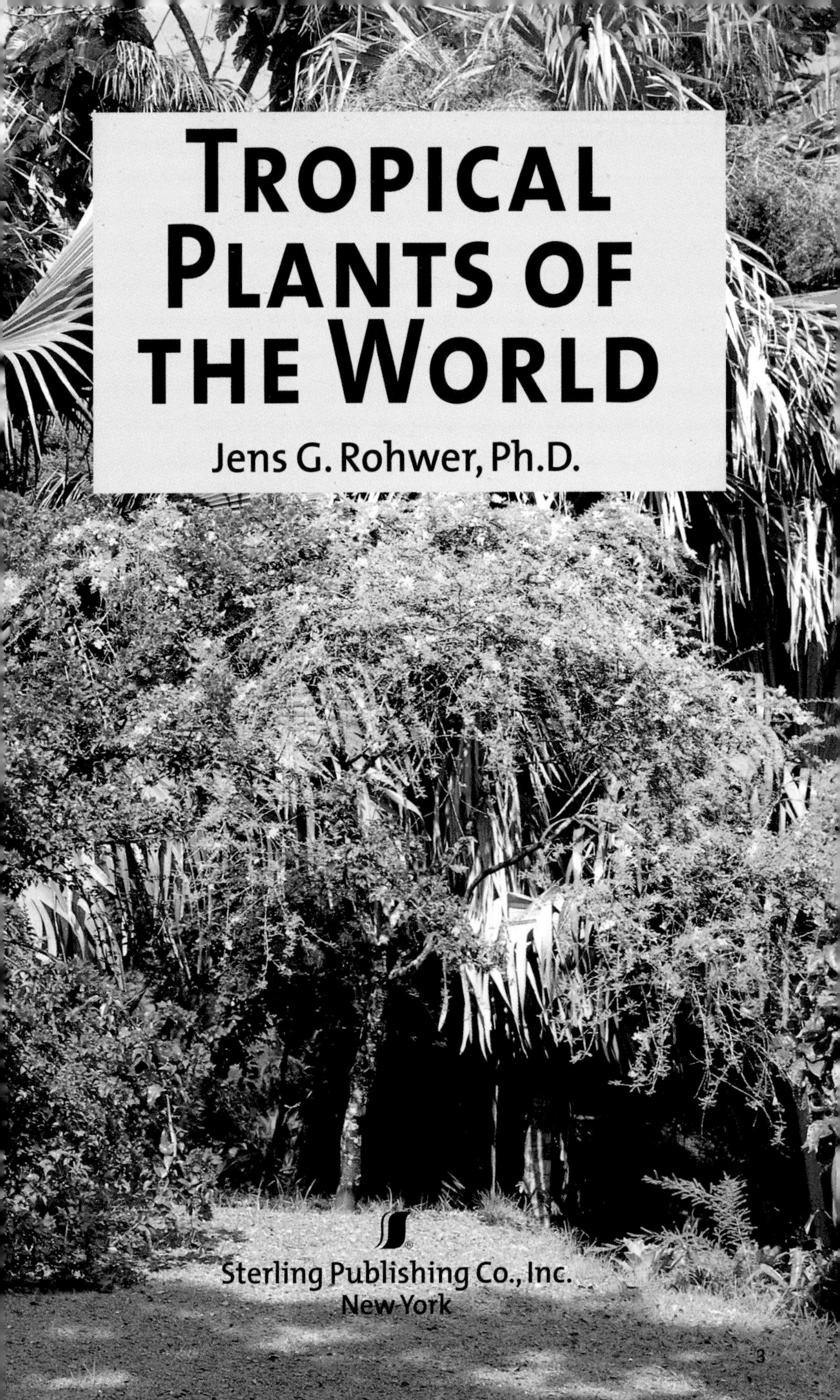
TROPICAL PLANTS OF THE WORLD
Jens G. Rohwer, Ph.D.
Sterling Publishing Co., Inc.
New-York
3

Contents

Introduction

The phrase *the tropics* conjures up images of blistering sun, shimmering midday heat, and short, intense thundershowers followed by a stifling mist. The phrase also brings to mind warm nights filled with the sound of cicadas, the buzzing of mosquitoes, seductive fragrances, and exotic palates. Above all, it suggests an overwhelming variety of life. No one is capable of counting the fauna in the tropics, which is home to millions of insects alone. On the other hand, we *only* have to deal with approximately 200,000 plant species, a number far greater than one person can categorize, let alone sum up in one book. Is the title *Tropical Plants of the World* somehow presumptuous?

Most tropical flora are so rare that relatively few people have the chance to catch even a glimpse of them. Many other species are more common but are inconspicuous in their forest habitat; thus, hardly any traveler will pay attention to them. Only a limited number of species live in accessible areas and are conspicuous enough for the casual observer to notice. This book is dedicated to those kinds of plants.

Obviously, the selection of species is always subjective. In general, I've included the ones that I paid attention to on my trips to South America and southern Asia. A comparison with other books of this kind will show that most authors include the same spectrum of species, no matter if the focus of their tropical experience is Latin America, Africa, Asia, or Hawaii. The reason for this limited range is simple: Almost from the beginning of time, man has transplanted decorative and useful plants from their place of origin to other areas. Now, these plants have spread all over the tropics. In some cases, scientists have a difficult time determining the origins of a plant.

Many ornamental plants give our living rooms a touch of tropical atmosphere. However, due to the relative lack of light and humidity, these plants seldom reach their full splendor indoors.

The Tropics

Although most people have a general idea of what the phrase *the tropics* means, several precise definitions are important. In the strictest geographical sense, the Tropics refers to the area between the two belts that encircle the earth: the Tropic of Cancer, located at 23°27' north latitude, and the Tropic of Capricorn, located at 23°27' south latitude. In this area, the sun is directly overhead at least once, and often twice a year. The difference between the summer months and the winter months is barely noticeable. Even on the shortest day, the tropical sun shines for ten and a half hours. It shines for thirteen and a half hours on the longest day. Most areas of North America experience a much greater fluctuation of daylight hours between summer and winter.

Notice that this strict definition includes areas of high mountains as well as desert areas. On the other hand, this definition excludes some areas in which the vegetation strikes one as being tropical, such as the southernmost region of Florida, the south of Brazil, Assam, and most of Taiwan.

Another definition revolves around the question of whether the average daily temperature fluctuation is greater than the average fluctuation over the course of the year. Basically, the question is whether the difference between day and night is greater than the difference between summer and winter. This definition extends the limit of the tropics to lowlands, where the influence of the ocean weakens the effect of the seasons, and to deserts and mountains, where the daily fluctuations are extremely wide.

Often the tropics are identified with summerlike heat all year round. This defini-

Many people think that beaches, palm trees, and coral reefs are typical of the tropics. Australia, Great Barrier Reef.

tion is correct if one regards the average yearly temperature as 78 to 82°F (26 to 28°C) at sea level. In this sense, all areas in which the temperatures never sink below the freezing point or, even in the coldest months, below 65°F (18°C) at sea level belong to the tropics. As the elevation increases, however, this definition loses its value because temperatures fall an average of about 3°F per 1,000 feet (2°C per 300 m).

Another definition relies on the wind and weather systems of the earth. In the equatorial area, humid air is constantly rising. As it rises, it cools and loses its humidity. The humidity turns to rain and merges with the rising heat. This often results in ferocious thunderstorms in the afternoon when the heat is most intense. The cooled air streams towards its respective pole until it sinks once more at the tropical line (23°27' north or south). At the same time, it becomes warmer and, therefore, drier. As a result, areas of low precipitation form. These have high air pressure all year round. These are the desert belts of the earth. From these belts, the air streams into the lower strata of the atmosphere and back to the equator, where it absorbs humidity once more as it crosses the oceans. Thus, the cycle starts anew.

Additionally, the rotation of the earth causes the streaming air to drift from the north-south axis; therefore, in the Northern Hemisphere, we have the northeast trade winds. In the Southern Hemisphere, we have the southeast trade winds. Although there may be a delay of up to two months, the center of this streaming system, with its heavy rains and constant low air pressure, follows the seasonal shift of the highest position of the sun to the north and to the south. According to this definition, only the area in this zone, which is hit by heavy rains at least once a year, can be regarded as the tropics.

The heavy rain more or less fulfills the needs of the flora; and, in fact, the amount of rain is one of the most important factors for vegetation. However, convergence is not the only source of precipitation in the tropics. The trade winds, which are generally rather dry, can also bring humidity with them, especially if they have covered large distances across the oceans. The rains appear when the trade winds cool down as they are forced to rise at mountain ranges. In East Asia, yet another variable adds to this cycle. There, the tropical low-pressure zone in summer shifts far into the Northern Hemisphere. As soon as the southeast trade winds have passed the equator, they drift in a westerly direction and result in the southwestern monsoon that brings a significant amount of humidity with it because of its long trip across the Indian Ocean. Similar, but less extensive, effects bring West and Central Africa (in the Southern Hemisphere) the northwestern monsoon and bring Australia additional rain showers.

Tropical Biomes

A biome is a major ecological division, such as a desert. Conditions in the tropics are very different from what we are used to in moderate latitudes. Almost constant temperatures and relatively stable length of days permit year-round growth of plants as long as enough water is available. However, the lack of cold in the winter months makes plants more vulnerable to pests. The amount of solar radiation is more than twice as high as it is in more moderate climates, and the same applies to the rate of evaporation. The latter extracts so much water from plants that only meadows could grow in the tropics if they had only the precipitation available in more moderate climates.

Soil conditions are also of special importance to the flora. Soil in the Northern Hemisphere appeared after the last ice age and is less than 10,000 years old. In the tropics, many landscape surfaces have been exposed to weathering for millions of years. Weathering in the tropics is essentially more intensive than in the moderate latitudes. Often the nutrient-binding minerals of clay were destroyed long ago, and rain washed out the nutrients. What remains is the typical red, tropical soil that, when dried out, can turn into rock-hard laterite.

The luxuriant vegetation of these soils remains alive only through constant recycling. When the dead parts of plants rot, actinomycetes absorb the nutrients before they can enter the soil. Thus, the plants receive nutrients from the bacteria rather than from the soil. If this cycle is interrupted, many nutrients are lost for good. Young soils, rich in nutrients, are found in the tropics in meadows close to rivers and in places where volcanoes have deposited their lava. Ironically, agriculture flourishes best in places where people are seriously threatened by flooding and volcanic eruptions.

Rain Forests

Although rain forests exist in moderate latitudes, many people think of the rain forest as the epitome of a tropical biome. Tropical rain forests were once the largest closed tracts of forest. However, in the second half of the twentieth century, developers and loggers destroyed many of these areas. More than half have already been irretrievably lost, along with hundreds of species of flora and fauna whose features we will never learn. Yet each year, an area approximately the size of Wyoming falls victim to the industrial desire for paper and plywood and to the enormous population pressures in the tropical countries themselves.

The tropical rain forest is not a homogeneous biome. Different geographic regions vary in soil, topography, precipitation, and distribution.

Continuously humid rain forests need more than 80 inches (2,000 mm) of precipitation per year, if possible, equally distributed; but they must have more than 4 inches (100 mm) monthly. In some loca-

Khao-Yai National Park, a rain forest in Thailand.

tions, precipitation exceeds 400 inches (10,000 mm) per year. Because the area above the treetops cools down at night, so much dew is created in the early morning hours that the lower levels of the forest remain very wet, even on rainless days. Such forests are extremely rich in flora, though the majority of species that root in the soil are trees. For example, in the Malaysian part of the Malaysian peninsula alone, scientists have identified more than three thousand species of trees whose trunks can reach a diameter of more than 12 inches (30 cm). By way of comparison, northern Europe has only about fifty species.

Occasionally, more than three hundred species of trees can be found on less than 3 acres (1 hectare); however, only a few examples of each species usually grow in these areas.

In Southeast Asia, one family of plants, the dipterous fruit plant (Dipterocarpaceae) dominates the great trees of almost every forest. The easiest way to identify these is the characteristic fruit (see photo at right). Because there are no seasons, these trees, which grow in humid rain forests, bloom and sprout at any time throughout the year. However, many of these trees grow 80 to 120 feet (24 to 36 m) high. Sometimes, single trees jut out far above the closed treetops, growing up to 230 feet (70 m) high.

One or two layers of shadow-bearing trees, high bushes, and the young growth of

Fruits of the dipterous fruit tree, Dipterocarpus bandii.

the tall trees grow below the closed treetops, waiting for a gap to appear in the canopy. Because little light reaches the ground, bottom growth is rare. Actually, the humid rain forest is not an impenetrable jungle. Instead, it resembles a cathedral with high rising pillars. Large amounts of life remain hidden on the ground from the observer. Only in open strips can the visitor view the variety.

Seasonal rain forests receive almost the same amount of precipitation as the continuously humid ones, but seasonal rain forests have to survive a dry period lasting for about a month. This regularly occurring dry period helps to regulate the trees so that the phases of blooming, fruit bearing, shedding of leaves, and new growth adjust themselves. Very few trees are completely without leaves in the dry period, and many are flowering during this time. The remaining species shed their old leaves after the growth of new ones or even carry several years of leaves.

In the seasonal rain forest, more light reaches the forest ground, increasing the number of leafy plants located there.

Deciduous forest near El Dorado in Venezuela.

Deciduous Forests

Even without human interference, gaps appear in rain forests. Wind, lightning, fungal infestation, and insect infestation all cause gaps. The death of an old tree rarely causes a real gap because the treetops of neighboring trees immediately spread into the space. On slopes, landslides create gaps; and in the dry forests, fire becomes a major factor. Rain forests are normally too wet to burn; however, when a fire does start, it can be devastating, as was the case in Southeast Asia in 1997 and 1998.

The ultimate destiny of a gap in the forest depends on its size. As mentioned above, neighboring treetops and young growth quickly fill in small gaps. In larger gaps, light-loving pioneer species can take root. These are species that do not appear in closed forests, but their seedlings are better able to compete than the seedlings of the forest trees. A deciduous forest (see photo at left) develops that is lower and has fewer species than the rain forest does, but the deciduous forest can show a larger variety of flora and fauna. Grasses often turn this type of forest into an impenetrable mass that includes varieties of bamboo, herbs, high shrubs, bushes, and climbing plants with thorns, all of which grow as tall as some trees. The most commonly appearing trees are the fast-growing pioneer trees with light wood, such as balsa trees (*Ochroma lagopus*), which grow more than 65 feet (20 m) high in only fifteen years. Some of these trees, such as *Cecropia* in South America and *Macaranga* (see photo at right) in Southeast Asia, live in a symbiotic relationship with ant colonies. These offer the ants hiding places in hollow branches and food in the form of special pods. In return, the ants attack anything that touches their tree, even human beings.

Within a few years, a deciduous forest becomes so dense that seedlings of timber forest species can only appear if seeds reach it in sufficient number. Sometimes, the original forest can regenerate itself within a few decades if the loss of nutrients was not too great during the gap phase. Farmers often use these clearings until the nutrients are completely exhausted; then, it is far more difficult for the forest to renew itself.

Mountainous Forests

Typically, rain forests at lower elevations are more luxuriant than those which develop at higher levels. Although precipitation is usually the same at higher elevations, the soil has no trapped moisture. This lack hinders growth because the surplus water can easily run off. In addition, the temperature decreases with increasing height, resulting in the condensation of humidity and the creation of clouds, fog, and dew. The height at which this takes place is variable; it depends on the initial values of the temperature, the humidity, as well as the flow of air. Slopes facing into the prevailing winds are much moister than those on the corresponding protected side because of the constant humidity.

In general, the size of the trees decreases as the height of the forest increases. At the same time, the forest has fewer woody vines, more epiphytes, and more moss and ferns, among them tree ferns which are suggestive of palm trees (see page 28). Typically, real palm trees, as well as bananas and a number of other tropical trees, remain below 3,300 feet (1,000 m). Some cultured plants, such as coffee, thrive at medium altitudes.

At higher levels, sometimes referred to as the cloud level, the temperature is cooler than it would be at corresponding altitudes due to the dense cloud cover. The relative humidity almost always remains at one hundred percent, so everything is wet, even if the rain gauge shows no precipitation.

***Macaranga triloba*, a tree typical of the deciduous forest, is probably the most common plant found in Southeast Asia. The food pods are situated below the shell-shaped minor leaves.**

In this misty forest, the trees are deformed and short, often with small, firm leaves. Moss covers everything, even the smallest twigs. Together with the constant fog, the moss gives this forest an eerie appearance. The bizarre species of this "elf forest" have adjusted themselves so closely to their biome that they are not able to live anywhere else. For this reason, they are not included in this book.

High Mountains

The border of trees in tropical latitudes follows the line of regularly occurring frosts. The line is usually between 10,000 and 13,000 feet (3,000 to 4,000 m). In some locales, it may be higher or lower. In the area above 13,000 feet (4000 m), the temperature drops below the freezing point almost every night; but during the day, the sun warms up the soil.

The phrase, "every day summer; every night winter," is only half true. Plants do not have to withstand long cold spells; they only have to endure a few hours. On the other hand, they still have to deal with extreme temperature fluctuations within a day. Species of high-mountain plants were not included in this book.

Seasonal Forests

Large areas in the tropics, especially those affected by the monsoons and the trade winds, show a distinct change between wet and dry periods. Where rainy periods which produce sufficient water are interrupted by two to five months of dry weather, one finds a mixed forest of evergreen and deciduous trees. The percentage depends on the humidity. In Sri Lanka, India, Burma, and Thailand, monsoon forests cover areas larger than the true rain forests. The trade wind forests on the Pacific side of Central America and the largest part of the tropical forests in northern Australia are considered seasonal forests.

Tillandsia ionantha, **a small orchid.**

Seasonal forests are not as rich in flora and fauna as rain forests. However, they appear to have a larger variety because different plants can adapt their growth, blooming and bearing fruit in response to the different conditions. Many species bloom during dry periods, which, fortunately for plant lovers, are also the best times to visit. Thus, the seasonal forests are among the most attractive biomes for visitors. These brighter forests support much more underbrush than do drier or more humid forests. In drier areas, one finds many grasses and herbs; in more humid areas, one sees shrubs and large-leafed climbing plants. From the start, vines have a better chance in seasonal forests than they do in the dark, closed rain forest. Epiphytes catch one's eye more easily, as they are not lost in the treetops, frequently settling on the trunks near the ground.

As a result of human interference, seasonal forests are the most damaged and endangered ecosystems within the tropics. They offer relatively favorable conditions for agriculture and, therefore, have been largely populated for a long time. In addition, they have a large share of valuable tropical wood such as teak (*Tectona grandis*, see page 92) and mahogany (*Swietenia* in America and *Khaya* in Africa), which the lumber industry has exploited. As a result of this exploitation, Africa no longer has a significant number of this type of forest, despite the fact that the climate in the vast areas north, south, and east of the Congo basin is ideal.

Dry Forest

While 20 to 80 inches (500 to 2000 mm) of rain fall within four to seven months, harsh conditions prevail during the remaining months, and most of the evergreen underbrush recedes. What results is dry, bare, and often only a few feet (meters) high. In areas with extensive rain, such as in large areas of Central America and Southeast Asia, one can often find a small mosaic of seasonal forests on the exposed sides and dry forests on the protected sides of the mountains.

During the rainy period, the treetops of these dry forests are dense enough to provide large areas of shade. On the other hand, the treetops allow enough light to support the luxuriant development of grasses and herbs. Late in the dry period, this forest appears dead; not only are the trees bare, but hardly any green remains at tree level. Only bushes with hard leaves, such as the succulents (cacti and similar fleshy plants), remain green. These plants have saved enough water in their stems or leaves to withstand the dry period. Yet before it starts to rain again, many trees start to bloom and bud. The new leaves are mostly thin and soft compared to those of the rain forest, and they are much less protected against dehydration. This obvious difference is easy to explain: The leaves of the rain forest must survive for several years, while the ones of the dry forest only survive during the wettest months of the year.

Where precipitation amounts to less than 20 inches (500 mm) per year or the length of the dry period lasts for more than eight months, the changes in the dry forest depend upon its location. Irregular rainfall and penetrable soil support the growth of thorny bushes and succulents without any measurable underbrush. This is the case in the southwest of Madagascar and in the northeastern section of Brazil. On the other hand, less penetrable soil and relatively regular precipitation, which occurs during the warmer season, give grasses an advantage over trees, resulting in grassy landscapes.

Plants from these very dry areas have only been included here if they are also frequently cultivated in wetter areas.

Savannas and Other Grassy Landscapes

Landscapes dominated by grasses or similar plants can develop in the tropics under completely different conditions. Small, reliable amounts of precipitation are only one of the possibilities. Very flat ground, fire during the dry period, and accumulated precipitation during the rainy period can also lead to open landscapes. In Africa, these cover extensive areas north, south, and east of the equatorial forest region. With increasing amounts of precipitation, a transition from plain grassland to thorny-bush savanna to tree savanna occurs. The thorny acacia (see page 70) and the baobab (*Adansonia digitata*, see page 84), which is able to save large amounts of water in its barrel-shaped trunk, are characteristic of this biome.

Mangroves

Anywhere along tropical coasts where the water is warm enough, where the coast is flat, and where the land is protected from strong breaking waves, one may find mangroves. Under the proper conditions, this vegetation reaches far above the tropical border. Mangroves usually grow in copses. Depending on the location, the plants are bushes or trees. Because the silt is soft and not well oxygenated, plants require special adaptations. Most species have prop or respiratory roots that grow out of the soil. These are roots that rise out of the soil or that grow out of the soil and then back in again. Roots exposed to low tides absorb oxygen from the air through their numerous pores. During high tides, the root system subsists on this oxygen. The numerous roots help

A typical mangrove woods with the prop roots that prevent erosion along flat coastlines.

collect and hold additional deposits of soil. In this manner, the mangrove belt protects areas of the coast from erosion and may even slowly extend itself towards the ocean.

However, with the increasing destruction of mangrove forests for tanning substances and charcoal, as well as for shrimp farms, the protection of the coast itself is endangered in many areas.

As an adaptation to the soft soil, some plants have become viviparous, meaning that the single seeds of each fruit germinate on the mother plant. These may grow 8 to 16 inches (20 to 40 cm) long. Sometimes they even grow up to 3 feet (1 m) in length. Typically, these seeds grow in the shape of an arrowhead.

When the seed grows too long for the plant to support it, it falls into the mucky soil. It may become stuck and continue to grow on that spot. More frequently, however, the seed drifts on the tide for a while before it washes ashore at another place. When this occurs, it quickly anchors itself.

Plant Species

Palms and Palmlike Plants

Botanists consider palms to be the quintessential tropical family. Indeed, their characteristic silhouettes create the picture of many tropical biomes. However, palms grow far beyond the tropics. One can find them in southern Europe, Japan, and New Zealand.

Most palms are easily recognizable: They have a slender, almost uniformly thick trunk with no branches. The top of a palm is a leafy crown of very large blades. The blades may be feathered (feather palms) or fanlike leaf blades (fan palms). These features are not always present. Among the more than 2,600 species are trunkless palms, palms with branches, climbing palms, and even species with relatively small leaves. On the other hand, not every plant that looks like a palm is a palm. In the tropics, one also finds tree-shaped ferns (usually in humid mountain forests) and the strange palm ferns which, despite their misleading name, are not related to palms or to ferns. In addition, some representatives of other species, such as the papaya (see page 52), can look like palms.

Tree ferns are spore plants. They do not reproduce by creating any blossoms or fruits. Instead, they reproduce by means of microscopic spores created on the underside of the leaf. To the naked eye, these spore reservoirs look like dark spots or lines on the leaves. The feathered leaves, called fronds, can reach a length of 7 feet (2 m). The youngest fronds in the center of the leafy crown are curled. As they mature, they uncoil. Unlike other trees, tree ferns do not produce wood. They owe the firmness of their trunks to a dense coat of short, firm roots streaked by the remaining stems of old fronds.

This book includes one of each of the two most important plant genera (*Cyathea* and *Dicksonia*, see page 28). They are typical of about 800 species of more or less tree-shaped ferns.

Palm ferns are regarded as "living fossils." The approximately 150 varieties are the last survivors of a much more richly developed group from the Proterozoic era. Their leaves are also very large, but they are feathered, and their young leaves are not curled. When the leaves are mature, they are usually stiff. Often they are sharp or stabbing. In addition, palm ferns do not produce any real blossoms. Instead, their male and female reproductive organs are situated in rather massive cones on separate male and female plants. As with some conifers, the pollen reservoirs are not enclosed in an ovary but are attached to the scales of the cone. In Cycadaceae (see page 30), the frondlike carpels are easily accessible.

Real palms, on the other hand, are flowering plants. Usually, they have numerous small flowers in often remarkably large inflorescences. Young palm leaves are never curled. Typically, they are erect before they unfold. As is the case with tree ferns and palm ferns, real palms are different from other trees in that their trunks are limited in circumference. Young plants remain short until they have reached their final circumference. Only after this has occurred do they grow in height.

Though common in man-made landscapes, palms are not often found in nature. Their popularity is partly due to their elegance. However, palms also belong to the group of useful plants of the tropics. Many species have edible fruits, and these can offer an essential contribution to nutrition. People weave baskets and robes from palm leaves or from their fibers. In many rural areas, people still cover their houses with palm leaves. Other material uses require the destruction of the entire tree. For example, in order to harvest sago from the bark of the sago palm trunk (see *Cycas*, page 30 and *Metroxylon sagu*, page 40), the whole tree must be cut down. Harvesting cabbage palm or palm hearts also kills the trees because it removes the vegetation point, the only spot from which palms can grow. These trees cannot branch out sideways or from below as other trees can. The wood of palms is sometimes used as firewood or for construction. Rattan furniture uses the slender and elastic trunks of climbing rattan palms (see photo, page 18). A listing of all these uses would fill many books, as would the description of the approximately 2,700 species. Therefore, we've included only those species which would be most likely to catch the eye of the tropical tourist.

Trees

In the tropics, the number of species is far higher than in moderate latitudes. In addition, the variety of growth structures and the splendor of the many flowers are overwhelming. Tropical trees owe both facts to the lack of winter. This is the only reason that the blossoming time of a single species can spread out throughout the year. The constant presence of nectar from their blossoms is a source of nutrients for larger animals, such as birds and bats. Strong pollinators capable of long flights are advantageous because plants from any one species are often separated by long distances. Of course, bigger animals need larger blossoms with more nectar than do smaller insects. In addition, birds often have a poor sense of smell. Thus, the blossoms must rely on the splendor of their colors to attract the necessary pollinators.

In moderate latitudes, winter also limits the variety of growth structures because it enforces dormancy and rest, usually with leaf shedding. In addition, trees in moderate latitudes cannot rely on only one or even a ,few places of vegetation (i.e. where buds grow); the risk of damage from one deadly frost would be much too high. Thick, water-saving trunks and leaves would be blasted

Buttress roots are characteristic of many trees in rain forests. Bukit Cahaya, Malaysia.

As with *Brownea ariza*, leaf shedding is a common phenomenon in the tropics.

by the frost and the wide-spreading, flat crowns or fan leaves would collapse under the burden of snow.

In the tropics, these kinds of dangers are missing. Because frost only occurs high in the mountains, the only rest phase that occurs is the result of a dry period. Therefore, the amount of precipitation or the plant itself regulates the rhythm of growth in continuously humid areas. Actually, continuous growth is rare because any tree that always has young leaves would soon be discovered as a reliable source of nutrients, assuming that it is not poisonous.

One rarely finds conifers in the tropics. They are most often found in mountain forests. Many of them have leaflike, widened needles or other growth structures that are unfamiliar to us. As a result, they are seldom recognized as coniferous. Only some Araucariaceae (see page 56) are frequently used in gardens.

With tropical deciduous plants, one can observe a number of plant structures that do not exist in moderate climates. Most conspicuous are the buttress roots in large trees. Because tropical soils are continuously humid and often saturated with water, they are poorly aerated, and the nutrients maintain a superficial cycle. Thus, most trees have a rather flat root system that may be insufficient to anchor them. Buttress roots offer a solution; they help to distribute the gripping power among as many surface roots as possible in order to resist wind.

Although aerial roots are characteristic of epiphytes (see pages 18 and 246–258), numerous trees, especially the many varieties of fig trees, have them. In some cases, aerial roots can grow considerable trunks and send far-reaching branches over long distances (see page 120).

Unlike deciduous trees in moderate latitudes, in the tropics, red-leafed deciduous leaves do not turn red shortly before they drop. Instead, the red color is actually a fairly typical characteristic of just-sprouted, young leaves. The green of the leaf develops later. Even more impressive is the procedure of leaf shedding. In this system, whole branches, including the leaves, sprout very quickly and hang down limply at first. Because they are slightly red or brownish in this stage, they appear almost as if they have been broken off. Later, however, the branches become erect, and then the leaves turn green.

Numerous tropical trees carry their sprouts and fruits on the trunk, on strong branches, or on short stocky branches. In some cases, as with the *Artocarpus integer* (see page 118), the thin branches of the crown would not be strong enough to carry the enormous weight of the fruits. In other cases, as with the *Parmentiera aculeata* (see page 86), this arrangement serves as the location of sprouts and helps direct the approach for pollinating bats. The long axes of inflorescences serve the same purpose; the blossoms project out of the top or protrude below it.

Bushes

Because trees dominate the humid tropics, two very different biomes exist for bushes. One involves the shade of the forest, and the other involves deciduous growth (see page 10). Usually, bushes that grow deep in the forest have either insignificant flowers, which rely completely on their fragrance to attract pollinators, or very bright, almost luminous flowers, which one can even see in the greenish twilight.

However, in the case of deciduous growth, many different plants are competing for space, light, and pollinators. Many have very remarkable flowers and are interesting as ornamental plants.

Creepers and Climbing Plants

One of the most remarkable features of tropical vegetation is the great number and variety of climbing plants. They can overgrow fences and bushes, wrap whole houses and tree trunks, hang in wide loops in forests, or even grow erect in order to reach treetop level. Curtains of vines, hiding everything behind them, are typical in places in which natural catastrophes have occurred or in which human beings have cut strips into the forest. On the other hand, in undisturbed forests, vines usually grow in isolation. They use a wide variety of methods to reach the light. Climbing plants, called root climbers, use their supports to climb. Winding climbers, on the other hand, wind their shoots around their support, as do plants such as string beans.

Tendril climbers, such as peas, use special, string-shaped, sensitive organs called tendrils. The tendrils remain oblong and move in circles until they find support. Then, they firmly wind themselves around the support. Most furl themselves like coiled springs with an elastic suspension. Spreading climbers grow through the trunk system of other plants and prevent themselves from slipping down with far-reaching branches, leaves, thorns, or burrs. Because they have to follow the movement of their supports during storms, vines need especially firm and elastic stems. For this reason, even the largest, woody vines rarely develop a regular wooden body as trees do. Where this wooden growth does develop along the vine, it can lead to bizarrely shaped stems.

One often sees large, isolated, hanging vines in the forest, and one wonders how they got up there, growing from tree to tree. The answer is fairly simple. Vines usually start their growth in gaps. They grow with the new forest up to the new height. In the treetops where the branches intersect, they can grow from tree to tree. However, because pioneer woods that die after a few years tend to dominate the gaps at first, the vines may lose the largest part of their original supports. If they have managed to anchor themselves also in longer-living plants, they can hang freely between these. In such cases, they often cover large distances. For example, scientists have measured rattan palms (see photo below) with a total length of 790 feet (240 m). They have measured *Bauhinia* that are more than 1,970 feet (600 m) long. Of course, vines of such length need a very effective water system in their stems. In fact, if one cuts some vines open, water will run out of them.

Herbs and Herbaceous Plants

Under natural conditions, herbaceous plants are not as prevalent in the humid tropics as they are in moderate latitudes. Like bushes, they are easily pushed aside into deep shadow. Often, they find room to grow in disturbed locations. Because wherever people live and work, they constantly disturb the forest, even in the tropics one finds a variety of flowering herbaceous plants. Firm montane perennial herbs, such as ginger and banana plants (see pages 50, 222, 224, 230, 232, and 238) can sometimes reach treelike dimensions. Most of the time, they grow in free fields or in the light forest.

On the other hand, in the closed forest, leafy plants often have strikingly colored or sculptured leaves. As much as this may look like a game of nature, it is important for the survival of the plants. The red colorings in the leaves help plants to use the energy-rich blue light, which is able to reach the ground. Plants need this blue light because the chlorophyll in the tallest trees has already absorbed the red light, which normally starts the process of photosynthesis. Strongly sculptured leaves improve the rate of evaporation, helping to carry nutrients through the plants, though this is difficult in the humid air on the ground in the forest.

Rattan palms at a water reservoir. Bukit Cahaya, Malaysia.

Epiphytes

Epiphytes are plants that grow on other plants. Epiphytes differ from parasites in that they do not penetrate into their host in order to withdraw water and nutrients. Instead, they use them only as a surface. If they damage their hosts at all, it is because both plants are competing for space and light. In addition, both plants are trying to catch water and nutrients before they reach the ground. Finally, epiphytes may harm their hosts by overloading the branches of their host with their own weight.

Some species of trees, such as *Saraca indica* (see page 64), are particularly attractive to epiphytes and always seem to be densely populated with them.

Because of the way they exist, epiphytes

This strangler fig has already wound around the trunk of its host.

This rain tree has many different epiphytes. San Fernando de Atabapo, Venezuela.

are able to reach the light using only a small amount of nutrients and energy. At the same time, they face some serious disadvantages. Unlike herbs, which live on the ground in the forest, epiphytes may be directly exposed to weather factors. They have to withstand strong winds, hefty rain showers, cold nighttime temperatures, and the burning heat of the sun.

Because they are not connected to the ground, their bark contains little water. As a result, epiphytes have to handle extreme dryness in continuously humid rain forests. Many species have developed water-saving tissues in their leaves or sprouts; some also have cisterns (see page 254).

As with most plants, epiphytes absorb water through their roots. In many cases, these have turned into aerial roots with special, external textures to absorb water. For example, the Bromeliaceae (see page 258) have developed water-absorbing scales on their leaves.

Most epiphytes are herbaceous. They are usually able to grow on the ground if they can successfully compete for light. On the other hand, ground-dwelling herbs can sometimes also grow in the forks of branches where enough humus has collected.

Some plants start as vines. Later, they lose contact with the ground and continue growing as epiphytes. On the other hand, some woody plants, such as figs (see page 120) and the autograph trees (see page 90), reverse this pattern of growth.

These plants start as small, epiphytic shrubs that then send aerial roots towards the ground. When the roots reach the ground, the plant can grow to forty times its size within a year and can virtually strangle its host (see photograph above).

The pineapple is a typical tropical plant cultivated on huge fields in many regions.

Marsh and Water Plants

Because of the high level of precipitation, the ground in the central tropics is often marshy. As a result, many plants actually grow or stand in water. In the Amazon River basin, and, to a lesser extent, in the areas around the Congo and Mekong Rivers, the soil in whole forests is flooded up to eight months per year. However, no one would refer to the trees that grow in these areas as water plants. Only plants which grow exclusively in water and which can survive only short dry periods are regarded as real marsh and water plants.

These plants have a larger capacity for growth than land plants because dry periods only change the water level. The large volume of water moderates the fluctuations of air temperature, and water birds easily spread the plants' seeds. Depending on the quality of the water, most of these plants can grow from their own fragments. Many species grow so rapidly that they create floating islands or even clog up whole bodies of water (see page 260).

However, most of these plants are rather insignificant or are only seldom seen. Therefore, this book only includes a few species which are either cultivated or are so widespread that one can hardly miss them.

Crops

In the tropics, agriculture occupies large areas. For example, from a plane, one can notice that oil palm plantations in Malaysia or rice fields in Thailand stretch over many hundreds of square miles (kilometers). The immense variety of flora in the tropics is reflected in agriculture; the number of species grown is well over a thousand. Even if most of these species are only locally important, many crops have regional or even international importance. In fact, the number of species is more than a book of this size can cover. For this reason, this book will discuss only a small selection of striking species. Fortunately, most people are already familiar with common species, such as rice or pineapple.

Plant Names

Most tropical plants don't have English names, any more than plants growing in North America have African or Asian names. However, the species shown in this book are different. They already have English names (often more than one) given by florists or gardeners who did not expect their customers to use unwieldy scientific names.

Often these names developed from the German, French, or Spanish names applied during the colonial period. Many of these folk names are not very specific. For example, if one wanders across the country and asks for "buttercup" or a "butterfly

Rice fields dominate the landscape in large areas of tropical Asia. Near Chom Tong, Thailand.

orchid," one can be sure to be offered a dozen different species for each name.

The only reliable source for further information is the scientific name, which is written in *italics*. The scientific name has two parts. The first one is the genus name, which is capitalized. The second, which is written in lower-case letters, indicates a specific species. The genus is the broad category that most amateurs would recognize. An oak tree, for example, belongs to the genus *Quercus*. Few amateurs would care whether they were dealing with *Quercus alba*, the white oak, or *Quercus rubur*, the northern red oak.

Plants that belong to the same species have essentially the same features. Usually, they are able to reproduce with each other and to create descendants that look like their parents. The exact definition of such a union is extremely controversial.

Of course, not all individuals of the same species are the same. Within a species, depending on the magnitude of the difference, one may also differentiate between subspecies, varieties, or shapes. In addition, the familiar name often has special significance for useful and ornamental plants. For example, when it comes to apples, names such as Golden Delicious or Granny Smith are very specific.

Above the level of genus, there are five levels of categories that describe a plant. Of these, this book only uses the level of family.

Tips for Using This Book

Categories

In order to help you find certain species, we've organized the plants in this book into eight categories:

Palmlike Plants (pages 28–46)

Trees (pages 48–120)

Bushes (pages 122–178)

Climbing Plants (pages 180–210)

Herbs and Herbaceous Plants (pages 212–244)

Epiphytes (pages 246–258)

Marsh and Water Plants (pages 260–264)

Crops (pages 266–274)

The section on **palmlike plants** includes tree ferns, palm ferns, and real palms. These are plants with only one leafy crown per trunk and either feathered or fanlike leaf blades.

Feather Palms

Fan Palms

On the other hand, you'll find giant-rosette plants with different leaf shapes at the beginning of the **tree** section (pages 48–52). Trees have one trunk and one top. This section includes species with slender, undivided, striped, oblong leaves with names such as yucca palm (see page 48) or seashore screw pine (see page 52), although they have nothing in common with palms.

Giant-rosette plants

Bushes also have woody parts, but they branch out from the ground.

Climbing plants can be herbaceous or have woody members, but they need support in order to continue to grow vertically. Without any support, they grow in a drooping or climbing manner.

Herbs and herbaceous plants do not create any woody parts, even though some of them can become quite large.

For the last three categories, the distinguishing characteristic is not the way they grow; it is their location that distinguishes them.

Epiphytes are plants that often grow on trees, but they are also kept as ornamental plants on the ground.

Unfortunately, the borders between the categories are not always very clear. Many species, such as those used for hedges, can become bushy if they are cut frequently or if they are gnawed or burnt when young.

On the other hand, as vines age, they can develop stems so thick that they are able to hold themselves erect. In these cases, only the twisted or bowed form of the stem allows us to determine its origin.

Many bushes have elastic, drooping branches, which can attach themselves to other bushes in the dense underbrush, creating a transition to a climbing organism. On the other hand, large vines can remain herbaceous if they do not find an adequate support or are cut regularly. Sometimes, the same species is treated in different ways. In South America, for example, people let bougainvillea (see page 198) climb like a vine. However, in Thailand and Malaysia, the fashion is to trim bougainvillea into a little tree or round bush.

In addition, the difference between bushes and herbaceous plants is not as clear in the tropics as it is in moderate latitudes. For instance, in moderate climates, when a plant buds after winter, it is a bush. However, if it freezes and dies down to the ground, it is considered to be an herbaceous plant. Of course, this kind of differentiation does not work in the tropics. The definition depends on the degree of woody material in the plant.

Subcategories and order of species

The larger categories are subdivided, usually according to the shape and order of their leaves (that is to say pinnately divided or palmately divided, simple, opposite, or in whorls, as well as simple and alternating).
Within these subcategories, the plants are arranged according to their colors, from blue to violet to red, orange, yellow, white, and green.

Leaves divided pinnately

Leaves divided palmately

Leaves divided pinnately or digitately arranged

Simple, opposite leaves

Simple, alternating leaves

This method of ordering, however, has its hidden weaknesses. For example, many species have large color varieties with multi-colored flowers or even flowers that change color during the flowering season. Some of the bracts in the inflorescences have a different color that is even more striking than the flowers themselves.
Sometimes leaves have transitions between lobed or deeply incised shapes on the one hand, and real pinnate or palmate leaves on the other (compare page 24). In addition, at first glance, it is not always easy to differentiate between a pinnule and a delicate lateral shoot with small leaves. A simple rule may help: Buds from which new branches or flowers can grow always develop in the axis (the point at which the petiole is attached to the main stem). The location of buds or lateral shoots, therefore, tells where the leaf starts.

Assignment of Photos

The photo appearing on the upper half of a page belongs to the listing opposite it. The photo on the lower half of the page belongs to the listing opposite it. If a photo illustrates something other than the listing opposite it, the "additional information" section will indicate the reference.

Glossary

Alternating: An arrangement of leaves in which each leaf is attached to the stem without a second leaf directly opposite it (Illustration 21).

Bilabiate: A corolla with a clearly distinguished labrum and labium (Illustration 3).

Bipinnate: A pinnate leaf with separate small stalks that have leaflets (Illustration 3).

Bract: A specialized leaf from which a lateral shoot, an inflorescence, or a flower grows. Usually used in connection with flowers, it is not actually a part of the flower (Illustrations 1 and 2). A flower bract looks different from a normal deciduous leaf. Often, it has strikingly colored leaves in the inflorescence.

Branch whorl: A circular arrangement of leaves, flowers, etc., formed on the stem at the same height (Illustration 24).

Capitulum: An inflorescence with many sessile flowers on a common main axis, such as is the case with the sunflower (Illustration 7).

Capsule: A fruit with more than one carpel, which opens when ripe.

Cob: An inflorescence with a thickened main axis and many small sessile flowers, as is the case with corn (Illustration 10).

Corolla lobes, cusps: The free part of a stunted whorl of petals (Illustration 3).

Corolla tube: The connected part of a stunted whorl of petals (Illustration 3).

Crown shaft: On palms, the green formation at the top of the trunk. Although it seems to be a continuation of the trunk, it is actually formed of intertwined boat-shaped bracts of palm fronds.

Dentate: Having small teethlike or, at least, blunt incisions between (Illustration 18). Similar to being serrated.

Digitately arranged: Having several small leaves (like the fingers of a hand) starting from the end of a petiole or leaf stalk (Illustration 14).

Distichous: A stem with leaves that alternate left and right with an angle of 180 degrees between the leaves (Illustration 22).

Epicalyx: An additional hull composed of small leaves around the calyx. Only a few species have an epicalyx (Illustration 2).

False umbel: A multiply branched-off inflorescence with the flowers located at essentially one level or situated spherically (Illustration 9).

Giant-rosette plants: A tree without branches that has a top of large leaves at the end of the trunk, such as virtually all palms.

Hermaphroditic: Plants whose flowers have both male and female reproductive organs (Illustrations 1, 2, and 3).

Inferior: An ovary situated below the other organs of the flower. It can be seen when looked at from the bottom of the flower (Illustration 3).

Labium: The bottom of a corolla in which the upper and lower parts can be clearly distinguished (Illustration 3).

Labrum: The upper part of a corolla in which the upper and lower sides can be clearly distinguished (Illustration 3).

Leaf blade: The area of the leaf (as differentiated from the stem of the leaf).

Obovate: Egg-shaped with the smaller, pointed end attached to the stem.

Opposite: Paired leaves that are attached to the stem opposite each other (Illustration 23).

Ovary: The lower part of the female reproduction organ of the flower (Illustrations 2 and 3).

Ovate: Egg-shaped with the wider, rounded end attached to the stem.

Palmately cleft: Incompletely separated lobes that start from the midrib (Illustration 16).

Palmately lobed: Divided lobes that are not completely separated. If the lobes start from one point, the leaf is **digitate or palmately**

lobed, like a maple leaf (Illustration 15). If the lobes start from the midrib, the leaf is **palmately cleft** (see previous definition).

Panicle: A multiply branched-out inflorescence. The grape is actually a grape panicle (Illustration 8).

Peltate: A leaf that is attached to the stem from its bottom or lower side instead of from its base, as is the case with the nasturtium.

Perianth: The outside organs of the flower, encompassing the stamens and carpels, basically the calyx and the corolla (Illustration 2).

Petal: One of the leaves of the corolla, often strikingly colored (Illustrations 2 and 3).

Petiole: The main stem of a leaf.

Pinna: The part of a pinnately divided leaf that branches off the midrib; it can be a pinnule or a lateral shoot of the first order (Illustrations 12 and 13).

Pinnately divided: Divided leaves with a stem-like midrib and lateral ribs or small leaflets arranged opposite each other (Illustration 12). When the small leaves first are attached to lateral ribs (pinna of the first order), the leaf is **bipinnately divided** (Illustration 13). In order to differentiate between pinnules and delicate branches with small leaves, see page 23.

Pistil: The female reproductive organ of the flower that starts from the ovary and continues through the stigma (Illustrations 2 and 3).

Raceme: See "simple raceme."

Reticulate venation: Lateral veins that branch off the midrib like the feathers of a bird. Most leaves show such venation (Illustration 11).

Serrated: Formed like the teeth of a saw (Illustration 17).

Sessile: Attached without a petiole.

Simple raceme: An inflorescence with one main axis (with no additional branches) from which flowers grow on petioles (Illustration 5).

Simple umbel: An inflorescence from which many petioles start from one point (Illustration 6).

Sinuated: Leaf edges with rounded shapes that approach and retreat from the middle several times. If the rounded shapes become larger and deeper, the leaves are lobed like oak leaves (Illustration 20).

Spike: An oblong inflorescence with stalkless flowers (Illustration 4).

Stigma: The part of the female reproductive organs which receives the pollen (Illustrations 1, 2, and 3).

Stipules: Leaflike organs at the bottom of the petiole. They are usually paired and often small (Illustration 11).

Subshrub: Borderline case between a bush and an herbaceous plant. Initially, it will have herbaceous branches, but later it will have small woody branches.

Superior: An ovary situated above the other organs of the flower, meaning that it is visible when looking from above into the flower (Illustration 2).

Three-numbered (also four-numbered, etc.)**:** In terms of leaves, having three small leaves on one common petiole (with only three leaves, pinnately divided and digitate can barely be differentiated). In terms of flowers, having three of all organs, for example trisepalous, tripetalous, six stamens, and a trilocular ovary. Although stamens often grow in doubled numbers, the ovary usually does not.

Unisexual: A flower with the reproductive organs of only one gender. (Most flowers have both a stamen, the male reproductive organ, and an ovary, the female reproductive organs, which makes them hermaphroditic.)

Winged: Having wide, flat edges. Winged petioles usually have two such edges; winged stems usually have four.

Species

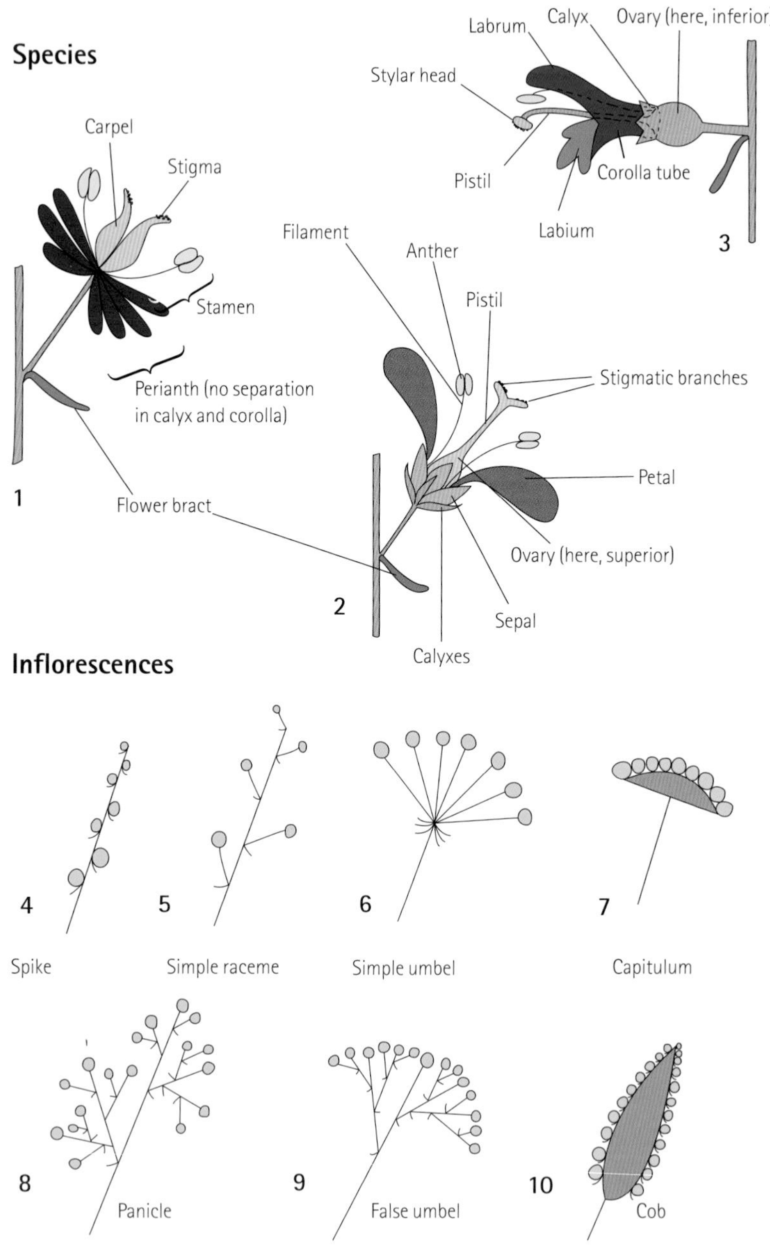

Labrum
Calyx
Ovary (here, inferior)
Stylar head
Carpel
Stigma
Pistil
Corolla tube
Labium
3
Filament
Anther
Pistil
Stamen
Perianth (no separation in calyx and corolla)
Stigmatic branches
Petal
1
Flower bract
Ovary (here, superior)
2
Sepal
Calyxes
Inflorescences
4
5
6
7
Spike
Simple raceme
Simple umbel
Capitulum
8
9
10
Panicle
False umbel
Cob

Leaf Parts and Leaf Margins

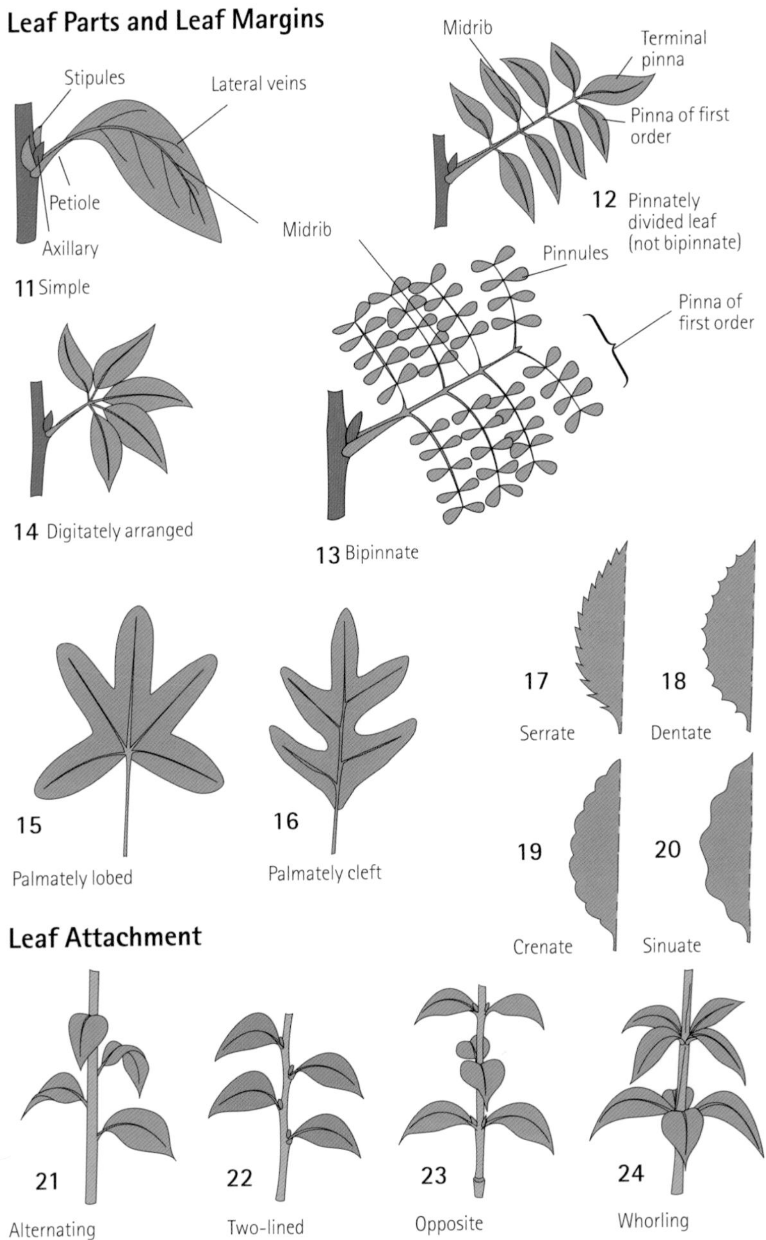

11 Simple

12 Pinnately divided leaf (not bipinnate)

13 Bipinnate

14 Digitately arranged

15 Palmately lobed

16 Palmately cleft

17 Serrate

18 Dentate

19 Crenate

20 Sinuate

Leaf Attachment

21 Alternating

22 Two-lined

23 Opposite

24 Whorling

Tasmanian Tree Fern

Dicksonia antarctica
Family: Dicksonia, *Dicksoniaceae*

Most important features: The strong trunk is densely covered with remnants of old petioles between numerous, short roots. The upper part of the trunk and the lower part of the fronds are densely covered with long, brown hair. The fronds are bipinnately or triple-pinnately feathered. Young fronds are curled snail-like and densely furred without scales or thorns. The spore reservoirs are located on the edge of the bottom side of the pinnules.
Growth structure: The Tasmanian tree fern grows straight, up to 50 feet (15 m) high, but most specimens never reach this height.
Leaves: The leaves are from 7 to 13 feet (2 to 4 m) long. At the beginning, they rise at an acute angle. The leaves are sessile pinnules. The longest are located in the middle of the frond. They are 12 to 16 inches (30 to 40 cm). The leaves have deeply incised edges that merge toward the pointed end.
Spore reservoirs: These are noticeable as yellowish or brown pods on the bottom side of the leaf, next to the margin.
Occurrence: This tree fern is typically found above 3,300 feet (1,000 m) in the southern or central tropics. Originally, it was from Tasmania.
Other name: German "Antarktischer Baumfarn."
Additional information: After Cyathea (see below), Dicksoniaceae is the second most important group of tree ferns, but it only includes twenty-five species. These are widely spread in the mountains and the moderate latitudes of the Southern Hemisphere. However, they reach as far north as Mexico and the Philippines. Marattia plants (Marattiaceae) have some treelike forms, yet they have short and thick trunks. These plants have fronds that grow up to 23 feet (7 m) long. Marattia plants are easily recognizable because of two shell-like formations to the left and right of the frond bases.

Australian Tree Fern

Cyathea australis
Family: Cyathea, *Cyatheaceae*

Most important features: The trunk is slender and densely populated with large, elliptical leaf scars. The top appears thickened because of the old leaf bases and the short roots. In younger plants, the whole trunk may look this way. The young fronds are curled, snail-like, and densely covered with brown scales.
Growth structure: The trunk grows straight, up to 65 feet (20 m) high. Most of these tree ferns are much smaller.
Leaves: The leaves are about 5 to 12 feet (1.5 to 3.5 m) long and extend like an umbrella. The leaves are bipinnately up to triple-pinnately divided. The pinnules are slender and oblong. Often, they are deeply cut on the edge. The leaves and pinnules are always widely separated toward the base but merge toward the tapered tip. The frondlike petioles and midribs are covered with scales and hair. Often, they also have bumps and thorns.
Spore reservoirs: The spore reservoirs are noticeable as spherical or kidney-shaped points on the bottom of the leaf. They appear on both sides of the midrib. Spore reservoirs are located on the bottom side of the pinnules, not on the edges.
Occurrence: One finds the Australian tree fern in tropical to moderate latitudes worldwide, especially in the area of humid mountain forests. It originated in Australia and Tasmania.
Other name: German "Australischer Baumfarn."
Additional information: The genus *Cyathea* includes about 650 very similar species, which grow naturally in only a limited area. Approximately half of the species have a membrane over the young spore reservoirs. Others, such as *Cyathea australis*, do not have this membrane. In earlier times, those were improperly categorized as *Alsophila*.

Cycad

Cycas circinalis
Family: *Cycadaceae*, real palm ferns

Most important features: The trunk is usually thick and short with acute scales. The leaves are pinnately divided and narrower at the base with distinctive midribs.
Growth structure: The trunk grows erect and usually is not branched. It seldom grows more than 10 feet (3 m) high.
Leaves: The leaves are 3 to 8 feet (1 to 2.5 m) long. The youngest leaves are curled. Older leaves are .4 to .6 inch (1 to 1.5 cm) wide. They are stiff but not poking. The order of pinnately divided leaves continues with thorns on the petioles.
Reproductive organs: With male plants, these occur as massive cones in the center of the leafy crown. With female plants, these are leaflike and light brown with round pollen reservoirs on the sides. They occur in batches, alternating with deciduous leaves at the top of the trunk.
Fruits: The fruits are orange red seeds about 7 to 10 feet (2 to 3 m) in diameter.
Occurrence: The cycad grows in all frost-free areas. Originally, it was from tropical Asia.
Other names: English "Fern Palm," "Sago Palm" (not to be confused with the real Sago Palm, page 40); German "Eingerollter Palmfarn."
Additional information: The seeds and young leaves are eaten cooked. Material from the core of the tree is used as a substitute for sago. However, one must exercise extreme care in handling any part of the cycad because all palm ferns contain highly effective nerve toxins. Among the seventeen species of the genus, the Japanese sago palm (*Cycas revoluta*) is often cultivated. It is originally from East Asia and differs from the others because of its narrow leaves and pinnas that are about .2 inch (5 mm) wide. These become smaller toward the leaf base. The difference in the carpels is even clearer. The cycad's carpels have a tapering, barely cut final section; the Japanese, on the other hand, have a wide and strongly tattered one.

Chestnut Dioon

Dioon edule
Family: *Zamiaceae*, Cone-palm Ferns

Most important features: The trunk is short and thick with long, pointed scales. The pinnas at the base are barely smaller than the ones in the middle. These have no midrib.
Growth structure: The chestnut dioon is erect, but almost trunkless. It never reaches a height of 7 feet (2 m).
Leaves: The leaves grow up to 6 feet (1.8 m) long. The youngest have pinnas folded above the midrib. Older pinnas are .2 inch (5 mm) wide. They are stiff with poking tips.
Reproductive organs: These grow in massive cones in the center of the frond crown. The cone scales have turned-up, pointed ends. The male cones have numerous pollen sacs on the bottom side; the female cones have two pollen reservoirs turned inward.
Fruits: The fruits are ovate, yellow seeds about .4 inch (10 mm) long. They are first visible when the orange red cone falls apart when it ripens.
Occurrence: One finds the chestnut dioon in many frost-free areas. Oiginally, it grew in Central America.
Other name: German "Mexikanischer Palmfarn."
Additional information: The seeds are rich in starch, which can be made into a type of flour. Some other palm ferns are cultivated. However, in contrast to this species, the pinnules at the bottom are remarkably smaller than in the middle. Further features include cone scales. In *Zamia*, they produce a hexagon pattern; in *Ceratozamia*, they have two thorns. With *Encephalartos*, they overlap, as is the case here; but they are blunter and usually produce rhombic patterns. In addition, most of the approximately thirty-five species have wider pinnules, which have at least some thorns on the side.

Peach Palm

Bactris gasipaes
Family: *Arecaceae*, Palms

Most important features: The trunk is densely covered with long thorns (although, sometimes only at the top) between narrow, thornless rings. Usually, several trees stand next to each other. The leaves are mostly yellowish green, and the tips of the pinnules are snapped off toward the ground.
Growth structure: This palm grows erect, up to about 65 feet (20 m) high; on rare occasions, one may grow 100 feet (30 m) high.
Leaves: The leaves are pinnately divided. They grow up to about 12 feet (3.5 m) long with a drooping arch shape. The petioles and midribs are thorny; the pinnas are very slender and grow in irregular groups.
Flowers: The flowers are yellowish white in branched inflorescences, that originate between the leaves. Later, however, they hang down. The bract whorls of the inflorescences are thorny.
Fruits: These are usually tricolored, ranging from green over yellow to orange red. They are about 2.4 inches (6 cm) long with three or six sides.
Occurrence: The peach palm grows in tropical America, but more precise information is unknown.
Other names: German "Pfirsichpalme"; Spanish "Chontas," "Pejibaye"; Portugese "Pupunha."
Additional Information: For many people of the Amazon basin and neighboring areas, the peach palms are among the most important and useful plants. After cooking for several hours, the pale yellow, mealy flesh of the fruit is edible. Because it contains great amounts of starch, it is suitable for flour and for fermenting the alcoholic beverages often prepared for ritual events. The extremely hard thorns of the trunk serve as pointed ends for the darts of blowguns. People use the fibers of petioles to weave robes. Many of the approximately 240 species of the genus *Bactris* are also useful.

Coconut Palm

Cocos nucifera
Family: *Arecaceae*, Palms

Most important features: The trunk is densely ringed with leaf scars. The very large and delicate leaves are pinnately divided and stand out in all directions. The fruits are very large.
Growth structure: The trunk rises in the shape of an arch and is seldom completely erect. The coconut palm grows up to 100 feet (30 m) high. When cultivated, it has a shorter, straight trunk.
Leaves: The leaves are up to 20 feet (6 m) long with a strong midrib and numerous narrow feathers up to 3 feet (1 m) long.
Flowers: These grow in large, yellow inflorescences in the center of the crown.
Fruits: The fruits are greenish to orange yellow and up to 12 inches (30 cm) long. They are oval and slightly angular.
Occurrence: Usually found naturally along coasts, these palms are often planted inland. Originally, they were from the western edge of the Pacific.
Other names: German "Kokospalme"; Spanish "Coco."
Additional information: The coconut palm is an extremely important plant because of its usefulness; no part remains unused. The trunk provides wood; the leaves, roofing and weaving material; the inflorescences, sweet juice for sugar and palm wine, and much more. Most important, however, are the fruits. The famous coconut is only the stone core. It is surrounded by a thick, fibrous cover with a leathery exterior shell. This gives the fruit its buoyancy. Fruits have been found after drifting for 2,800 miles (4,500 km) in the ocean over a four-month period. Inside, the nut contains a firm, white, sugary, edible fiber which can be made into dried coconut and coconut oil. In addition, younger nuts contain a clear, refreshing fluid called coconut milk. The latter is not actually in the nut; it develops when ground up edible fiber is filled with hot water, and the water is pressed out.

Oil Palm

Elaeis guineensis
Family: *Arecaceae*, Palms

Most important features: The trunk is firm and covered with old leaf spathes and fronds. The leaves are very large with firm thorns below the feathers. The inflorescences appear between the leaves. The flowers and fruits are densely packed.
Growth structure: The trunk is erect and grows up to 100 feet (30 m) high. When cultivated, it does not grow as tall.
Leaves: The leaves are pinnately divided and up to 23 feet (7 m) long. Most rise at the base and then grow in a wide drooping arch. The single pinnas are not equally distributed and are not all on one level.
Flowers: These are small and dirty white. They are densely pressed at the inflorescences, which end in a thorny acute tip.
Fruits: Most of the fruits are orange red. Often they are yellow at the base and almost black at the top. The shape is an oblong oval. Most have three sides that are up to 2 inches (5 cm) long. Hundreds of fruits grow in compact bunches that weigh up to 55 pounds (25 kg).
Occurrence: Oil palms are cultivated on enormous plantations in Southeast Asia. They are originally from West Africa.
Other name: German "Ölpalme."
Additional information: Oil palms are among the most important providers of raw material for the food and cosmetic industries. Palm oil comes from the fleshy fruit and is frequently used to produce margarine. The core holds palm curd oil, used in the production of soap. With up to 6 tons (5,400 kg) of oil per acre (.4 hectare) each year, oil palm plantations are among the most profitable industries in tropical cultures. It is hardly surprising that each year thousands of acres (hectares) of rain forest are destroyed to create new plantations. This is especially true in Malaysia, which produces approximately two-thirds of the world's supply of palm oil.

Pygmy Date Palm

Phoenix roebelenii
Family: *Arecaceae*, Palms

Most important features: These are delicate, small palms with V-shaped, folded leaves. The midribs face the bottom side, and the edges face the upper side of the leaf.
Growth structure: Pygmy date palms grow erect, up to a height of 8 feet (2.5 m). The trunk is densely covered with the remains of old leaves.
Leaves: These are delicately pinnately divided. They grow up to 3 feet (1m) long and are shaped like a drooping arch. The feathers are small and taper at the end.
Flowers: The flowers are small and yellowish. The inflorescences appear between the leaves.
Fruits: The fruits are black. They are elliptical to tapered and grow up to .4 inch (1cm) long.
Occurrence: This widely disseminated garden and pot plant is originally from Indochina.
Other names: English "Roebelen Date Palm"; German "Zwergdattelpalme."
Additional information: The date palms of the *Phoenix* genus are easily recognizable: The feathers of their leaves are folded in a V-shape and are pointed at the end. The small leaves of all other feather palms are either folded in an A-shape (with the midribs upward) or are not pointed. Two of the total seventeen species are cultivated more frequently than the pygmy date palm. However, these are in subtropical areas. Without the date palm (*Phoenix dactylifera*), life in many desert areas would be unbearable. The trunk of this palm exhibits a rhombus pattern created from the bases of old leaf bases. Usually, the relatively loose frond crown sits in the middle of twenty to forty leaves, each of which can reach a length of up to 16 feet (5 m). The Canary Island date palm (*Phoenix canariensis*) looks very much like the date palm. However, it has an even firmer trunk and a denser frond crown with up to one hundred leaves. Its fruits are smaller than real dates and have hardly any fleshy matter. They are considered inedible.

Betel Nut Palm

Areca catechu
Family: *Arecaceae*, Palms

Most important features: The trunk is slender with indistinct leaf scars. The leaves are relatively small and very densely pinnately divided. They often appear disorderly. The inflorescences and clumps of fruit grow below the crown, a little below the smooth, green shaft.
Growth structure: The betel-nut palm grows erect. Although it usually grows 33 to 50 feet (10 to 15 m) tall, it may reach a height of 100 feet (30 m).
Leaves: The leaves are pinnately divided. They are usually about 7 feet (2 m) long and stand out in all directions, drooping slightly at the tip. The leaflets are not completely equal in pinnate fronds; therefore, the single pinnas often have differing numbers of folds and several acute tips.
Flowers: These are small and light yellow. They grow in delicately branched-off inflorescences that look like powder puffs.
Fruits: The fruits are yellow to orange and spherical to ovate. They grow up to 2.4 inches (6 cm) in diameter.
Occurrence: The betel-nut palm is frequently cultivated in Southeast Asia. It was originally from the same area.
Other names: German "Betelpalme"; Malaysian "Pinang."
Additional information: The seeds of the betel-nut palm are considered a delicacy in Asia. Its nutritious fiber is cut into slices and wrapped together with calcium and spices in the leaves of the betel pepper (see page 268). When chewed, this snack frees the slightly numbing properties and the red dye. It also increases the flow of saliva and colors it red, probably the origin of some cannibal tales. Chewing betel nuts is supposed to suppress hunger and kill intestinal parasites; however, it also colors the teeth black and causes cancer of the mouth. Although people rarely chew betel nuts today, grooms in Thailand still give betel nuts and betel peppers to the parents of the bride.

Royal Palm

Roystonea regia
Family: Arecaceae, Palms

Most important features: The trunk is smooth and slender. The pinnas are not completely on one level. The inflorescences grow a little below the smooth, green crown shaft, which is slightly thickened at the base.
Growth structure: The royal palm grows erect, up to 80 feet (25 m) high. The trunk may have the same thickness from top to bottom.
Leaves: The leaves are pinnately divided, up to 12 feet (3.5 m) long. They rise steeply and them descend, droping slightly. The crown shaft is strikingly long, often up to 7 feet (2 m).
Flowers: These are yellowish white and grow in large, rich, branched-off inflorescences.
Fruits: The fruits are red, then black. They are almost spherical and about .4 inch (1 cm) in diameter.
Occurrence: Typically, one finds them in parks throughout the North American tropics. Often, they are cultivated to frame streets and other spaces. Originally, they were from Cuba.
Other name: German "Königspalme."
Additional information: Because it sheds its old leaves completely, the royal palm has a smooth trunk. This gives the appearance of a well-tended frond crown; on the other hand, since the leaves can weigh up to 45 pounds (20 kg), they pose a danger to pedestrians when they fall down. The presence of royal palms is an indication of good soil in the area. For this reason, almost all of its original area has been converted to agricultural purposes. Yet, it has not become rare; its elegant appearance makes it an attractive plant to cultivate. The West Indian cabbage palm (*Roystonea oleracea*), which can grow even taller and whose pinnately divided leaves stand off in clear order, is also often cultivated. The end with the youngest leaves is eaten like a cabbage (see page 44).

Bottle Palm

Hyophorbe lagenicaulis
Family: *Arecaceae*, Palms

Most important features: The trunk is short and barrel-shaped close to the base. It has ring-shaped leaf scars. The leafy crown is much thinner than the thickest part of the trunk., The frond crown usually has only five to eight leaves.
Growth structure: The bottle palm grows straight, up to 16 feet (5 m) high.
Leaves: The leaves are pinnately divided. Usually, they are less than 7 feet (2 m) long. The youngest are straight; the older leaves are arch-shaped and drooping. The leaf bases produce a relatively long, green, increasingly slender crown shaft.
Flowers: The flowers are small and yellowish. They grow just a little below the crown shaft in hanging inflorescences that look like powder puffs.
Fruits: These are orange to black spheres. They are about .8 inches (2 cm) long with well-preserved petals.
Occurrence: Bottle palms, usually found in closed parks and on hotel grounds, originally, came from Mauritius and a neighboring island.
Other names: German "Flaschenpalme"; French "Palmier Gargoulette."
Additional information: The trunk of a palm usually remains equally thick through its whole length. However, bottle palms of medium age develop distinctly barrel-shaped trunks. With age, the upper, narrower part becomes longer, which results in the bottle shape. Because of this bizarre form, its relatively small size, and its great tolerance for dryness and salt, the bottle palm has become a popular and expensive ornamental plant. The spindle palm, *Hyophorbe verschaffeltii*, is also very popular, but it grows much taller and has only a vaguely spindle-shaped trunk. All five species of the *Hyophorbe* genus have become extremely rare in nature. Only one specimen of *Hyophorbe amaricaulis* is still supposed to be alive.

Sealing-wax Palm

Cyrtostachys renda
Family: *Arecaceae*, Palms

Most important features: The slender, green trunk has ring-shaped gray leaf scars and bright orange to red leaf spathes and petioles.
Growth structure: The trunk grows erect, up to 40 feet (12 m) high; when cultivated, it is often shorter. The trees always stand together in dense groups. The younger trunks look like green bamboo canes; the older ones are gray.
Leaves: The leaves are delicately pinnately divided. They grow up to about 5 feet (1.5 m) long, usually rising until almost erect. They hardly droop.
Flowers: These are small and yellowish white. They grow in great numbers on long lateral shoots of inflorescences, which develop right below the crown shaft.
Fruits: The fruits are spherical to elliptic and about .4 inch (1 cm) in diameter. They are black when ripe.
Occurrence: The sealing-wax palm is popular all over the world as an ornamental palm. Originally, it was from the Malay peninsula, Sumatra, and Borneo.
Other names: German "Rotstielpalme," "Siegellackpalme"; Malayan "Pinang Rajah" (which means "royal palm").
Additional information: Cultivated sealing-wax palms only reach a height of 13 to 26 feet (4 to 8 m) and often have particularly striking red leaf sheathes. In earlier times, this color was used as a way to separate it as a proper species (*C. lakka*) from its larger and paler wild relatives; today, it is only considered a variation in size and color. Some other, more delicate, feather palms that grow in groups are often cultivated, such as parlor palms (*Chamaedorea spp.*) and the Madagascar palm (*Chrysalidocarpus lutescens*) illustrated on the right. Mountain palms normally become only thumb thick and often carry bizarre red clumps on their thin trunks. Golden leaf palms are stronger and have yellow green inflorescences and drooping leaves.

Fishtail Palm

Caryota mitis
Family: *Arecaceae*, Palms

Most important features: The leaves are bipinnately divided. The inflorescences and fruit clusters are distributed on the trunk, flowering or ripening from above to below.
Growth structure: The trunk is erect and grows up to 40 feet (12 m) high. It is slender, green or gray, and covered with leaf spathes. Most of the time several trees grow very close together.
Leaves: The leaves are up to 10 feet (3 m) long, with palmate fanlike leaves that look as if they are crooked triangles. At the tip, they appear to be cut off, pinnately divided leaves.
Flowers: The flowers are small and greenish. The inflorescences look like big powder puffs. On the branches that are sideways to the trunk, they look like a string of pearls.
Fruits: These are spherical and up to .8 inch (2 cm) in diameter. They remain green for a long time, then turn red brown and finally black.
Occurrence: Now found worldwide, it originally spread from India to the Philippines.
Other name: German "Milde Fischschwanzpalme."
Additional information: Fishtail palms are the only palms with bipinnately divided leaves. They grow without flowering until they have reached their final height. Then, inflorescences appear sideways on the trunk, first at the top, then farther and farther down, continuing toward the bottom. As the lowest fruits are ripening, the plant dies. The sugar palm (*Arenga pinnata*) has the same appearance; however, it has pinnately divided leaves. The Borneo fishtail palm (*Caryota no*) and the fire palm (*Caryota urens*) are widely spread. Both are larger than the fishtail palm. In addition, their leaves are more slender and are pinnately divided. The burning palm can cause itching when touched and burning in the mouth if the fruit is eaten. The fibers of its spathes are used to produce brushes and wickerwork. Their inflorescences are often tapped for the sugary juice, which is fermented into palm wine.

Sago Palm

Metroxylon sagu
Family: *Arecaceae*, Palms

Most important features: The thick trunk has wide, triangle-shaped remainders of old leaves at the top. The petioles are about half as long or longer than the pinnately divided part. Sometimes, the trees have dark rows with thorns. The inflorescences (when present) are located above the frond crown and are very large.
Growth structure: The sago palm grows erect, up to 65 feet (20 m) high; when cultivated, it is rarely more than 33 feet (10 m) high.
Leaves: The leaves are pinnately divided and grow up to 23 feet (7 m) long, rising steeply. Lower pinnules are not as stiff and are wider spread. The upper leaves are denser and more pointed toward the end.
Flowers: These grow up to .4 inch (1 cm) long. Many grow at the side axis of the second order of an enormous, regularly branched inflorescence at the top of the plant.
Fruits: The fruits are spherical to egg-shaped. They grow up to 3.2 inches (8 cm) long and are densely covered with yellowish brown scales.
Occurrence: Sago palms, found in Southeast Asia and the Pacific, usually in marshlands, originally came from Moluccas and New Guinea.
Other name: German "Sagopalme" (see also page 30).
Additional information: The sago palm grows rapidly and retains great amounts of starch in its trunk. As a result, it has become a dietary staple in many areas of the Western Pacific. People fell the trees, chop the cores, and rinse the starch out with water. If the starch pap and water are forced through a strainer onto hot metal, the result is pearl sago. The harvest occurs in cycles of ten years because each single trunk is rather short lived. After ten to fifteen years, the trunk flowers and dies soon afterward. However, the plant survives because of its basal shoots. People covet the leaves of the sago palm for roofing material because the leaves last for about seven years, longer than most other materials do.

Petticoat Palm

Copernicia macroglossa
Family: *Arecaceae*, Palms

Most important features: This is a small palm with short petioles and almost spherical leaves that are palmately divided. The oldest leaves continue to spread and to stand out almost horizontally. Below the crown, the trunk is covered with a dense coat of dead leaves.
Growth structure: The petticoat palm grows erect, usually only 13 to 16 feet (4 to 5 m) tall, but occasionally up to 26 feet (8 m) high.
Leaves: The leaves are palmately divided and up to about 7 feet (2 m) in diameter. The single fanlike leaf blades are only connected in the lower third of the leaf. The petioles are hardly recognizable. The "coat," which is made of dead leaves, is almost as wide as the crown. In taller specimens, it doesn't always reach the ground.
Flowers: These are small and greenish yellow. They grow in several branched inflorescences.
Fruits: The spherical fruits are almost .6 inch (1.5 cm) in diameter.
Occurrence: Petticoat palms are usually found in park grounds. Originally, they were from Cuba.
Other name: German "Breitfächrige Wachspalme" (see also "California Fan Palm").
Additional information: These palms have a thin layer of wax, which protects young leaves from drying out. Most of the twenty-nine *Copernicia* species are limited to Cuba. Only three kinds can be found in South America. There the carnauba wax palm (*Copernicia prunifera*) is widely spread. This is the most economically significant species because it supplies carnauba wax, used in lipsticks, shoe polish, and car polish. To extract the wax, the leaves are dried in the sun until the wax can be knocked off. More than one thousand young leaves are required to extract 2 pounds (1 kg) of wax.

California Fan Palm

Washingtonia filifera
Family: *Arecaceae*, Palms

Most important features: These are tall palms with long, stalked, palmately divided leaves. A dense coat of dead leaves covers the trunk.
Growth structure: The California fan palm grows erect, up to a height of about 50 feet (15 m). Without the old leaves (see photo at left), the palm appears completely different than it does with its leafy coat.
Leaves: These are palmately divided. The leaves, which are about 5 feet (1.5 m) long, stand out in all directions. Half to two-thirds of the length of the single fanlike leaf blades are connected, folded V-like, and split at the pointed ends. Often, they are partly dissolved into single fibers. The petioles are about as long as the leaf blades.
Flowers: The small flowers are pinkish white, growing in several branched inflorescences between the leaves and usually towering above.
Fruits: The fruits are almost spherical and up to .4 inch (1 cm) long.
Occurrence: Originally, they grew in the more humid spots of the dry areas of California and Arizona; now, they are often found in the humid tropics in parks.
Other names: German "Washingtonie"; English "Petticoat Palm" (also see "Petticoat Palm").
Additional information: Although the California fan palm is not a tropical plant, it is cultivated in the tropics for its unusual appearance. In humid areas, the leafy coat partly rots and tears off under its own weight. What remains is a relatively smooth trunk with muted leaf scars. In dry areas, on the other hand, the dead leaves are often removed by hand because they are highly flammable. A triangle-shaped pattern of old leaf bases is all that remains. As a plant growing relatively far north, the California fan palm can withstand slight frost; for this reason, it also thrives in the Mediterranean area.

Lontar Palm

Borassus flabellifer
Family: *Arecaceae*, Palms

Most important features: The remainder of old leave bases covers the top of the trunk. The bottom of the trunk is thicker than the top. The fan crown is almost spherical. Lower fanlike leaf blades are much shorter than the upper ones. The petioles are irregularly dentated.
Growth structure: The lontar palm grows erect, up to 100 feet (30 m) high. The trunk of older palms often varies in thickness. Sometimes, it is slightly swollen in the middle.
Leaves: The leaves are palmately divided, up to 10 feet (3 m) in diameter, with spherical to dragon-shaped outlines. About half of the single fanlike leaf blades are separated, folded V-like, and split at the pointed ends. The petioles are slightly longer than the fanlike leaf blades, clearly reaching into the leaf blade.
Flowers: The inflorescences are hidden between leaves and are very different in male and female plants.
Fruits: These are slightly flattened, spherical, and 4.8 to 8 inches (12 to 20 cm) in diameter. Usually, they have three pits with several scaly leaves at the base.
Occurrence: Lontar palms are widely spread in the tropics of North and South America. Originally, they spread from India to the Malay Peninsula.
Other names: English "Toddy Palm," "Wine Palm"; German "Palmyrapalme."
Additional information: In India, the lontar palm is the most important source of sugar juice from which sugar and palm wine are made. In fact, the lontar palm has more than eight hundred different uses, including providing wood to construct dams. Sometimes old specimens show a light swelling in the middle of the trunk. This is more distinct in the African *Borassus aethiopum*t. Doum palms (*Hyphaene*) appear to be similar, but they are smaller and easily distinguishable because of their regular, long thorns at the petioles.

Chinese Fan Palm

Livistona chinensis
Family: *Arecaceae*, Palms

Most important features: The Chinese fan palm has an almost spherical crown. The fanlike leaf blades are only connected in the lower one-fourth to one-third of the tree. After half to two-thirds of its length, it breaks off toward the ground, and the branches are threadlike and drooping.
Growth structure: The trunk is erect and grows up to 50 feet (15 m) high. It is grayish brown with blurred ring-shaped leaf scars. It may also have remainders of old leaves at the very top.
Leaves: These are palmately divided. Including the petioles, they are up to 13 feet (4 m) long. The petioles are always shorter than the leaf blades and usually about as long as the straight snapped-off parts.
Flowers: The flowers are yellowish and small. They grow in richly branched inflorescences, which remain hidden between the leaves.
Fruits: The fruits are blue green, elliptical, and about .8 inch (2 cm) long.
Occurrence: Chinese fan palms are usually found in parks. Originally, they were from a subtropical species that grew in South China and Japan.
Other name: German "Chinesische Livistonie."
Additional information: The genus *Livistona* is named after Baron Livistone, who founded the botanical garden in Edinburgh, Scotland, in 1670. It encompasses twenty-seven species, spread from Australia eastward to the Solomon Islands and northward up to the Himalayas and southern Japan. In addition, it includes one species from the Horn of Africa and one from Yemen. A related species, the Australian fan palm or cabbage palm (*Livistonia australis*) is often cultivated. It grows taller than its Chinese relative and has longer petioles, but its leaf blades are smaller. It owes its name to the fact that the pointed end of the shoot with the youngest leaves can be eaten as "palm cabbage" (compare with page 36).

Talipot Palm

Corypha umbraculifera
Family: *Arecaceae*, Palms

Most important features: The trunk is massive with ring-shaped leaf scars. The leaf blades are huge and fanlike. The inflorescence above the frond crown is even larger than the crown.
Growth structure: The palm grows erect, up to 80 feet (25 m) high. However, for many years, it is trunkless.
Leaves: The leaves are palmately divided with a spherical outline. They are up to 16 feet (5 m) in diameter. The single fanlike leaf blades are separated halfway, folded in a V-shape, and split at their acute tip. The petioles are very firm and grow up to 16 feet (5 m) long. The leaves of young plants rise, but during the blooming season, they hang.
Flowers: These are small and white to yellowish. Millions of them grow in an enormous, 20 to 26 feet (6 to 8 m) high, richly branched inflorescence at the top of the plant.
Fruits: The fruits are almost spherical and .8 to 1.6 inches (2 to 4 cm) in diameter.
Occurrence: Often, the talipot palm grows as a single specimen in parks. Originally from South Asia, these palms spread to South India, Sri Lanka, and Indonesia. The precise origin is unknown because it was cultivated very early in history.
Other name: German "Talipotpalme."
Additional information: In earlier times, the huge leaves of the talipot palm were used as umbrellas, as roofing, and as weaving and writing material. The plant only blooms once in its life, after fifty to sixty years. However, when it finally blooms, it produces the greatest inflorescences in the entire flora. The number of blossoms is estimated at 10 million. Although each blossom does not produce a fruit, about a year after the blossoms appear, 2 tons (3,600 kg) of fruits can ripen on one palm. With the ripening of the fruits, the plant uses its last reserves, and it dies. Sago, sugar, and fishtail palms (see page 40) bloom only once also.

Fan Palm

Licuala grandis
Family: *Arecaceae*, Palms

Most important features: This is a small palm with a thin trunk and great, long-stalked palmately divided leaves. The edge of the leaf is very regularly serrated. Each indentation is bipartite through a short cut.
Growth structure: The trunk grows straight, up to 7 to 10 feet (2 to 3 m) high. It is very thin and partly covered at the top with the remainders of old petioles.
Leaves: The leaves are palmately divided, more than half to three-fourths spherical, almost 3 feet (1 m) in diameter, and flatly expanded on the side. The fanlike leaf blades are only divided in the last few inches (cms) and bipartite at the end. The rachises grow up to 3 feet (1 m) long.
Flowers: These are yellow and about .4 inch (1 cm) long. They grow in multiply branched inflorescences, which bloom between the leaves and are only slightly larger than the leaves.
Fruits: The fruits are bright light red and about .4 inch (1 cm) in diameter. The clusters hang down.
Occurrence: One finds fan palms in gardens and parks in all humid tropical areas. Originally, they were from the island of New Britannia northeast of New Guinea.
Other name: German "Großblättrige Strahlenpalme."
Additional information: The fan palms are a rather large group with 108 known species. They grow from India to South China, Northern Australia, and the Pacific islands. Most of the species can be found in Borneo and New Guinea, usually in the brushwood of the rain forests. These fan palms are often smaller than the other species, and the leaves are usually separated into segments of several connected leaf blades (e.g. *Licuala spinosa*, see photo at right). The closely related (and often preferred) *L.peltata*, on the other hand, is not divided. This species has even larger leaves. It also differs from the former in almost vertical inflorescences that are only singly branched.

Bitter Aloe

Aloe ferox
Family: *Asphodelaceae*, Asphodill

Most important features: A mop of large, thick-fleshed, thorn-covered leaves covers the trunk. The flowers are orange to scarlet in one inflorescence with several straight spikes.
Growth structure: The bitter aloe grows up to 10 feet (3 m) tall, occasionally reaching a height of up to 16 feet (5 m) high. Typically, it is not branched. Dried leaves cover the trunk to the ground.
Leaves: The leaves are up to 3 feet (1 m) long and 6 inches (15 cm) thick and dull green. On the edges are sharp, reddish brown prickles. In addition, the leaf blades have sporadic prickles.
Flowers: The flowers grow in groups of six. They are up to 1.4 inches (3.5 cm) long and tubular. The pointed ends of the three inner petals are brown to black. The stamens reach far out of the tube. The inflorescence is up to 32 inches (80 cm) high.
Fruits: These are long capsules with three chambers.
Occurrence: Bitter aloe is found in parks and gardens. Originally, it was from South Africa.
Other name: German "Wilde Aloe."
Additional information: While bitter aloe is only used as an ornamental plant in other regions, in its place of origin, it has other uses. When still fresh, its viscous juice is used to produce gelatin. Its thorns offer protection from predators. If the thorns are removed, the leaves can serve as food for cattle. Many of the 370 species are cultivated. Most famous is the Barbados aloe (*A. vera*, see photo at right) with its yellow flowers and thorny leaves only on the edge. The juice from this species has long been used to heal wounds and to cool a sunburn. Today, it can be found in many cosmetics. The yellow juice offers some protection against ultraviolet light for the outer leaf layers, whereas the clear juice of the inner layers keeps the humidity in.

Elephant Yucca

Yucca guatemalensis
Family: *Agavaceae*, Agave

Most important features: The tuberous trunk thickens at the bottom. The branches are thick and erect. They have a mop of very large, tapered leaves at the end.
Growth structure: The elephant yucca grows up to 50 feet (15 m) high. Sometimes, it has multiple trunks with dichotomous branches and very dense, leafy tops.
Leaves: The leaves have sharp edges that taper to a slender point. They are 24 to 40 inches (60 to 100 cm) long and 2 to 4 inches (5 to 10 cm) wide. They are narrowed at the bottom, but they have no petiole. The youngest ones are erect; the oldest hang down.
Flowers: The flowers are cream white, hanging, and bell-shaped. They grow up to 3.2 inches (8 cm) long. Many grow together in an erect, flower panicle that is up to 35 inches (90 cm) high above the leafy top.
Fruits: The fruits are up to 4 inches (10 cm) long. They consist of capsules with three parts and many black seeds.
Occurrence: Elephant yucca is often found in gardens. Sometimes, it is also planted as a stockade or barrier. Originally, it spread from Mexico to Guatemala.
Other names: English "Spineless Yucca"; German "Riesenpalmlilie"; Spanish "Isote," "Itabo."
Additional information: The flowers of the elephant yucca are rich in vitamin C. They are eaten uncooked in salad or baked in pastries. For pollination, most elephant yuccas have to rely on certain moths, which transfer pollen and lay their eggs in the flowers. The grubs nurture themselves from a part of the growing seeds, but there remains enough to ensure the reproduction of the plant. Other, smaller species are often cultivated, especially the aloe yucca (*Y. aloifolia*), which has stiff, poking leaves and often a yellowish white edge. The Adam's needle (*Y. filamentosa*) can withstand winter temperatures.

Banana Tree

Musa X paradisiaca
Family: *Musaceae*, Banana plants

Most important features: The crown has very large, oblong leaves with clearly distinguishable petioles and strong, green midribs. The inflorescences and fruit clusters are drooping. The bananas appear at the base of the crown, and the flowers appear at the top.
Growth structure: These trees can grow up to 33 feet (10 m) tall; most species, however, are smaller.
Leaves: The alternating leaves are up to 13 feet (4 m) long. The lateral veins are almost vertical and stand out from the midrib. The leaves are pinnately incised.
Flowers: The inflorescences are up to 5 feet (1.5 m) long with large, dark red to brown violet whorls of bracts in the axis of which are ten to twenty pale, yellowish, tubular individual flowers.
Fruits: Each subspecies varies from green to orange red in color and from 4 to 15 inches (10 to 35 cm) in length. The fruits grow in groups at the base of the clusters, bent toward the top.
Occurrence: Banana trees are very widely spread. They originated in Southeast Asia.
Other names: German "Banane"; English "Plantain"; Malayen "Pisang"; Spanish "Plátano."
Additional information: The banana is one of the most important tropical foods and is even found in remote areas. This species has a variety of cultural types, from small, sweet baby bananas to large, mealy plantains. Actually, the banana is not a tree but a giant herbaceous plant, whose trunk consists of onion-shaped, interlocking leaf spathes. The abaca (*M. textilis*) provides Manila hemp used in homes and more delicate fibers used for tea bags and banknotes. The *Ensete ventricosum* is often cultivated as an ornamental banana. One can easily recognize it because of the sessile leaves and pink midribs.

Traveler's Tree

Ravenala madagascariensis
Family: *Strelitziaceae*, Strelitzia plants

Most important features: The very large leaves of the crown are palmately divided and spread out, fanlike, on one level.
Growth structure: The traveler's tree grows up to 100 feet (30 m) high. It begins growing without a trunk.
Leaves: The leaves are oblong, like banana leaves, 3 to 13 feet (1 to 4 m) long. They have long petioles and standing lateral views that are almost vertical from the midrib. In between, the leaves are often pinnately incised.
Flowers: The inflorescences grow between the petioles, up to 33 inches (85 cm) long, with five to fifteen firm, green, boat-shaped bracts. Each holds sixteen cream white flowers, up to 8 inches (20 cm) long. Two of the six petals are overgrown. Hidden among them are six stamens.
Fruits: The fruits have three chambers inside the woody capsules. The black seeds have a blue cover.
Occurrence: This is one of the most popular park trees. It originated in Madagascar.
Other names: German "Baum der Reisenden"; French "Arbre du Voyageur"; Spanish "Arbol de Viajero."
Additional information: The name of the tree refers to the legend that it is always a source of fresh drinking water for the traveler. Rain water collects in the axis of the firm petioles, which have a U-shaped cross section at the bottom. However, one must be close to dying of thirst in order to appreciate it. The collected water is full of moldering leaves, mosquito larvae, and similar unappetizing pests. The claim that the palmate fanlike leaf blades were always oriented in a north-south direction is nothing but a story. The unusually firm flowers are not able to open by themselves. In order to do so, they need the help of lemurs, the typical primates of Madagascar, which are rewarded with plenty of nectar for their assistance.

Papaya

Carica papaya
Family: *Caricaceae*, Papaya plants

Most important features: Most papaya trunks are not branched. They have a crown of huge, palmately cleft leaves that resemble giant fingers. Each of these leaves is deeply and multiply incised. The large fruits grow at the bottom of the crown. All parts have a milky sap.
Growth structure: Papayas grow up to 33 feet (10 m) high. They have a slender trunk, which is densely covered by leaf scars.
Leaves: The leaves accumulate at the top of the trunk. The petioles are up to 3 feet (1m) long. The leaf blade is only slightly shorter and deeply divided into five to nine lobes.
Flowers: These are cream white and mostly separated. The male flower (see photo at left) has a slender corolla tube in long, richly branched-off inflorescences. The female one is almost sessile and much thicker with only a very short tube.
Fruits: The fruits are greenish to orange yellow, shaped like a pear or melon, and up to 20 inches (50 cm) long. Usually, however, they are only about half that size. The flesh of the fruit is yellowish to orange red. The center consists of many black, slippery seeds.
Occurrence: The papaya is one of the most typical plants found in tropical gardens. Originally, it was from tropical America.
Other names: English "Pawpaw"; German "Melonenbaum"; Spanish "Lechosa"; Portuguese "Mamão."
Additional information: The papaya is more of a giant herb than a tree. Its hollow trunk develops only a little soft wood. The fruits are among the most popular of all the tropical food. They are picked before they are ripe for export; thus, the color is not very impressive. In Asia, unripe fruits are cooked as vegetables. The spicy seeds are sometimes used as a substitute for pepper. Papain, found in the milky sap is used, among other things, to promote digestion, soften meat, and prevent yarn from shrinking. Occasionally, one sees the mountain papaya (*C. pubescens*). It has hairy leaves and fruits with five sides.

Seashore Screw-pine

Pandanus tectorius
Family: *Pandanaceae*, Screw-pine plants

Most important features: This plant has many thick and branched prop roots. The leaves at the end of the branches grow in three screw-shaped and twisted sequences. The leaves are grasslike and oblong. Often they have sharp thorns on the edge and on the bottom side of the midrib.
Growth structure: Seashore screw-pines grow up to 33 feet (10 m) high. The trunk is light gray and ringed. It often divides into two or three almost equally strong branches.
Leaves: The leaves are 30 to 70 inches (80 to 180 cm) long. On occasion, they grow as long as 10 feet (3 m). The leaves are 1.6 to 3.6 inches (4 to 9 cm) wide, bluish green, very tough, and tapering toward the pointed tip.
Flowers: The male and female inflorescences grow on separate plants. The male flower can be up to 24 inches (60 cm) long with light bracts. A white spike can be found in each axis. The female inflorescences are spherical, about 2 inches (5 cm) long, and surrounded by whorls of bracts.
Fruits: The fruit clusters are spherical and up to 10 inches (25 cm) in diameter. The orange fruits grow in forty to eighty groups. Each group has five to eleven individual fruits.
Occurrence: Seashore screw-pines are widely cultivated. They originated along the coasts from Sri Lanka to Hawaii.
Other names: German "Duftender Schraubenbaum"; Malayan"Pandang"; Polynesian (Tahiti) "Fara"; Hawaiian "Hala"; Spanish "Palma de Cinta."
Additional information: Hawaiian girls regarded the pollen as a means of attracting young men. In addition, the fruits were considered a food staple on many Pacific islands. The variety of species is very large. Some species lack thorns on their leaves. In species with thorns, these have to be removed in order to make baskets and mats. Many of the approximately seven hundred *Pandanus* species are also useful. Some grow much larger.

Australian Pine

Casuarina equisetifolia
Family: *Casuarinaceae*, Pine plants

Most important features: The delicate, hanging branches are reminiscent of long pine needles. When examined closely, they resemble horsetails. The leaves are only tiny scales.
Growth structure: This tree grows up to 80 feet (25 m) high. The crown is very loose and delicately branched. The younger branches are green, delicate, and rippled. Some of the branches are partly shed as if they were leaves.
Leaves: These are arranged in whorls and are delicately dentate.
Flowers: The flowers are tiny. The male ones are up to 1.6 inch (4 cm) long with slender spikes at the end of the branches. The female ones (see photo at left) are less than .2 inch (5 mm) long. They have spherical capitula at short lateral shoots and red stigmas.
Fruits: These are spherical, woody cones up to .6 inch (15 mm) long with sharp-edged scales between tiny winged nuts.
Occurrence: The Australian pine grows on many tropical coasts. People cultivate it inland. As the name implies, it was originally found from Australia to South Asia and the Pacific.
Other names: English "Beach She-Oak," "Horsetail Tree," "Ironwood"; German "Strand-kasuarine"; Spanish "Pino Australiano."
Additional information: The Australian pine grows fast, endures salt and strong winds, and survives in even the poorest soil since actinomycete bacteria live in the small root tubercles and supply the tree with carbon dioxide from the air. Thus, this tree is ideal for protecting the coast and for reforesting eroded lands. In Florida, it is planted as a protective belt around lemon plantations. Its hard, heavy wood is excellent for firewood because it burns for a long time and is almost smoke-free. The Polynesians often produced weapons from it; in the red juice of the tree, they saw the blood of fallen warriors. They carefully dug out the roots, bent them, and returned them to the soil for a few more years in order to produce hooks for shark fishing.

Giant Bamboo

Dendrocalamus giganteus
Family: *Poaceae*, Grasses

Most important features: These grow as enormous, wooden bamboo canes. Many of them stand together, creating dense thickets. The trunks are often a dull green. The young shoots are cone-shaped with bluish black, triangular leaves on the edge (see photo at left).
Growth structure: Giant bamboo grow up to 115 feet (35 m) high and up to 12 inches (30 cm) in diameter. The branches are thin and delicately divided. Near the top, they stand sideways. Some of the lower joints of the trunk have roots.
Leaves: The leaves are oblong, up to 24 inches (60 cm) long, and 4 inches (10 cm) wide. They have delicate, parallel veins that run almost parallel to the edges. The leaves only appear near the top.
Flowers: These grow in large panicles with groups of .4 to .8 inch (12 to 20 mm) long spikes from which yellow anthers or violet stigmas hang. They are rarely seen.
Fruits: The fruits are like grains: they are only up to .3 inch (8 mm) long. Again, they are rarely seen.
Occurrence: One finds giant bamboo growing in parks. Originally, they were from Southeast Asia.
Other names: German "Riesenbambus"; French "Bambou Géant."
Additional information: Giant bamboo are actually the largest blades of grass in the world. They can grow in record time. For example, under favorable conditions, they can grow up to 18 inches (45 cm) per day. They are very stable with a hard, smooth surface. This fact makes giant bamboo very popular for constructing bridges, houses, and scaffolds. Because they are hollow between the joints, they are also suitable for rafts, tubes, musical instruments, and receptacles of all kind. Like almost all bamboo species, this one only blooms once in several decades and dies above ground after the fruits ripen. The young sprouts of many (such as the closely related *D. asper*) are eaten as vegetables.

Bunya Pine

Araucaria bidwilii
Family: *Araucariaceae*, Araucaria

Most important features: The bunya pine is a firlike tree with a more-or-less rounded crown. The branches stand out horizontally. Most of each branch is bare. Only the end of the branch has needles and needle-carrying lateral branches.
Growth structure: These pines grow erect, up to 165 feet (50 m) high.
Leaves: The leaves are needle-shaped, hard, and pointed. They are up to 2 inches (5 cm) long and .4 inch (1cm) wide. The young plants and branches bear small cones.
Flowers: The male cones are slender and up to 8 inches (20 cm) long. They fall off after the blooming season. The cones are smaller and oval in shape. They develop into large fruit cones.
Fruits: The fruit cones are oval, up to 12 inches (30 cm) long, and weigh up to 12 pounds (5 kg). They grow as densely connected, woody cone scales, each with a short, upwardly directed pointed end.
Occurrence: Bunya pines grow in parks and ornamental gardens. Originally, they were from Queensland in Australia.
Other name: German "Bunyatanne."
Additional information: Australian aborigines eat the seeds, which are up to 2 inches (5 cm) long, uncooked or roasted. The right to use a single tree is handed down from father to son. As a young plant, the bunya pine shows the typically cone-shaped coniferous structure; later, however, it develops from a wide, umbrella-shape into an almost flat crown. The name comes from the long, bent, branches with their wide, pointed needles that remain on the tree for many years. The bunya pine can be cultivated in areas with mild winter months. The monkey puzzle tree (*A. araucana*), often grows in tropical gardens, but it is not really a tropical plant. Originally, it was from the harsh southern point of South America.

Norfolk Island Pine

Araucaria heterophylla
Family: *Araucariaceae*, Araucaria

Most important features: This is a firlike tree with a cone-shaped crown. The branches grow in regular whorls. The oldest branches are arch-shaped and hanging; the younger ones are more erect. The very youngest branches are steeply erect.
Growth structure: Norfolk Island pines grow erect, up to 230 feet (70 m) high.
Leaves: The needle-shaped leaves are soft and up to .6 inch (1.5 cm) long in young plants. The leaves of older plants (especially those with cone-carrying branches) are much shorter. They are bent inside and have a hard tip.
Flowers: The male cones are long and slender. They fall off after blooming. The female cones are rounder, growing into large fruit cones.
Fruits: The fruit cones are spherical or slightly thicker than they are long. They grow up to 4.8 inches (12 cm) in diameter. They are covered by many densely connected cone scales, each with a thin, bent, pointed end about .4 inch (1 cm) long.
Occurrence: These pines are found in parks and ornamental gardens. Originally, they were from Norfolk Island in the Pacific Ocean.
Other names: German "Norfolktanne," "Zimmertanne."
Additional information: The Norfolk Island pine is one of the few conifers that thrive in the tropical lowlands. Today, they are cultivated as decorative trees; however, they became popular as the worldwide demand for sail boat masts increased. Because this pine grows rapidly and erectly, shipbuilders brought it to Hawaii, just as they did the very similar New Caledonia Pine (*A. columnaris*). Today, these species form a large part of the landscape in some places. Most of the eighteen *Araucaria* species are originally from the Southwest Pacific; two, however, come from the southern point of South America.

Jacaranda

Jacaranda mimosifolia
Family: *Bignoniaceae*, Trumpet tree plants

Most important features: The leaves are set opposite and are bipinnately divided. The flowers are light blue to light violet and shaped like thimbles. The trunk is short and light gray. It has a loose crown.
Growth structure: This tree grows up to 65 feet (20 m) tall.
Leaves: The leaves are approximately 8 to 16 inches (20 to 40 cm) long with up to twenty pinnately divided pairs. These are numerous, about .2 to .8 inch (0.5 to 2 cm) long, narrow, and delicately tapering. During the dry period, they fall off.
Flowers: The flowers appear with or just before the leaves sprout. They grow in panicles at the end of the branches. They are often lavender blue and up to 2 inches (5 cm) long. The corolla tube has five lobes and four stamens.
Fruits: These are light brown, oval to spherical, wooden capsules that grow up to 3.2 inches (8 cm) long. They are flattened on one side, often with wavy edges, and open between the two disk-shaped parts.
Occurrence: These trees are often cultivated in tropical and subtropical areas with alternating humidity. In places such as Hawaii and Zimbabwe, they grow wild. Originally, they were from southern Brazil and northwest Argentina.
Other names: English "Fern Tree"; German "Jacaranda"; Spanish "Guarupa."
Additional information: The jacaranda is one of the few plants whose folk name is the same in all languages. During the flowering period, the jacaranda offers a spectacular picture. The tree and the ground below it are completely covered with flowers. A few species have white flowers.

Sausage Tree

Kigelia africana
Family: *Bignoniaceae*, Trumpet tree plants

Most important features: The leaves are set opposite and are bipinnately divided. The flowers are very large and wine red to purple in color. The external portion of the flower has yellow veins. Gigantic, sausagelike fruits hang from the tree.
Growth structure: Sausage trees grow up to 60 feet (18 m) tall. At first, the bark is smooth and gray, and the crown , spherical. Older trees develop a scaly, peeling bark and a wide crown.
Leaves: The leaves grow opposite each other or in whorls of three. They are up to 20 inches (50 cm) long with seven to eleven pinnules; these may be up to 6 inches (15 cm) long. They are oval, stiff, and leathery with rough hair. The leaves fall off during the dry period.
Flowers: The flowers grow in hanging inflorescences that are up to 3 feet (1 m) long and up to 6 inches (15 cm) wide. The corollas are funnel-shaped and slightly bent with five rolled lobes. The corollas have four stamens inside.
Fruits: These are gray brown, up to 3 feet (1 m) long, and 7.2 inches (18 cm) thick. They weigh approximately 24 pounds (10 kg) and hang on long petioles that may be 3 feet (1 m) long.
Occurrence: These are ornamental trees not seen frequently. They are indigenous to humidity alternating areas of tropical and southern Africa.
Other names: German "Leberwurstbaum"; Spanish "Arbol Salchicha."
Additional information: Each flower of the sausage tree opens for only one night. The flower has an unpleasant fragrance, which attracts bats for pollination. The strange fruits, from which the tree gets its name, are not edible. However, people have found multiple uses for the fruits in folk medicine and magic. From rheumatism to snakebites and from syphilis to evil spirits and even to tornadoes, people believe the fruits help almost everything.

Flamboyant

Delonix regia
Family: *Caesalpiniaceae*, Carob plants

Most important features: The wide crown is shaped like an umbrella. The leaves are bipinnately divided. Large red flowers have petioles.
Growth structure: These trees may grow up to 50 feet (15 m) high, but they are often smaller.
Leaves: The leaves are alternating. They grow up to 20 inches (50 cm) long. They are fernlike and delicately divided with hundreds of pinnules, each about .4 inch (1 cm) long. The leaves can fall off during the dry period. However, this is only for a short period.
Flowers: The flowers are orange to scarlet red and 4 to 6 inches (10 to 15 cm) long with five sepals. The sepals are green on the outside and red inside. There are five petals. The upper part is often partly white or yellow with red spots. Inside are ten long filaments and one pistil. Most flowers appear at the end of the dry period together with new leaves.
Fruits: The fruits are black brown and flattened with a wooden hull. They are 12 to 24 inches (30 to 60 cm) long and 1.6 to 2.8 inches (4 to 7 cm) wide. They hang and are often slightly crooked and cross-ribbed. They remain on the tree until the flowering season.
Occurrence: Flamboyant is one of the most popular of all ornamental trees. Originally, it came from Madagascar.
Other names: English "Flame of the Forest," "Flame Tree," "Royal Poinciana"; German "Feuerbaum," "Flammenbaum"; Spanish "Arbol de Fuego," "Guacamaya," "Tabuchín."
Additional information: Flamboyant is its French name. Almost all languages have adopted the name. It is one of the most striking tropical trees. Although it started its triumphal procession around the world in the middle of the previous century, it now appears to be indigenous in many places. Day moths usually pollinate its flowers, but other insects and even birds also serve the same purpose. The tiny pinnules fold themselves together at dusk.

Orchid Tree

Amherstia nobilis
Family: *Caesalpiniaceae*, Carob plants

Most important features: The leaves are large and pinnately divided. The hanging inflorescences grow up to 3 feet (1 m) long. The bright red flowers are strikingly large with bizarre shapes.
Growth structure: The orchid tree grows up to 60 feet (18 m) high with a spherical crown.
Leaves: These are alternating, pinnately divided, and up to about 3 feet (1 m) long. Young branches and their leaves hang. They are slightly streaked with red at first. Later, they turn green and become erect, shedding their deciduous leaves.
Flowers: The flowers of the orchid tree are bright red or occasionally pink. They are 8 inches (20 cm) long and grow in hanging simple racemes that are up to 3 feet (1 m) long. The inflorescences have two bright red leaves that stand out. The flower itself has four smaller sepals and three larger petals. The latter have yellow spots near the top. The middle petal is wider than the two on the sides. Nine out of ten stamens grown into a tube; five of them are much longer than the tube and bend upward.
Fruits: The hulls of the fruit are very large. One seldom sees the fruits.
Occurrence: Typically, one finds these trees in Southeast Asia. Originally, they were from Burma.
Other names: English "Pride of Burma," "Flame Amherstia"; German "Tohabaum."
Additional information: In full bloom, the orchid tree is one of the most splendid of the tropical trees. Europeans first discovered it in 1826 in a temple garden. It has only been found twice growing wild. The orchid tree needs a lot of rain, high humidity, and rich soil. However, it does not do well with too much rain. The trees rarely produce fruits when cultivated. In addition, attempting to grow the tree from a cutting is not simple. Therefore, it is hardly surprising that, despite its splendor, the species is rather rare.

African Tulip Tree

Spathodea campanulata
Family: *Bignoniaceae*, Trumpet tree plants

Most important features: The leaves are pinnately divided and grow opposite each other or grow in three whorls. The flowers are very large, orange red, and bent. They grow in dense, erect, simple umbel-shaped racemes at the end of branches.
Growth structure: The tree grows up to 80 feet (25 m) high with a dense, somewhat oval top.
Leaves: The leaves grow up to 16 inches (40 cm) long with nine to twenty-one small leaves. The leaflets are dark green, elliptical to oval, and up to 6.4 inches (16 cm) long.
Flowers: The buds are brown, hairy, claw-shaped, and bent toward the middle of the inflorescence. They open from the edge. The corolla comes out sideways from the calyx. The corolla is up to 6 inches (15 cm) long and 3.2 inches (8 cm) wide with five lobes and thin, yellow, wavy limbs. Inside, it is yellow with red stripes. It has four stamens and a two-lobed pistil.
Fruits: The fruits are boat-shaped, up to 10 inches (25 cm) long, erect, and brown. When they are ripe, they split into two parts with many winged seeds.
Occurrence: One of the most frequently used ornamental plants, the tree is indigenous to tropical Africa.
Other names: English "Fire Tree," "Flame of the Forest," "Fountain Tree," "Nandi Flame"; German "Afrikanischer Tulpenbaum"; French "Immortel Étranger," "Bâton de Sorcier"; Spanish "Caoba de Santo Domingo," "Gallito," "Tulipán Africano," "llamarada de Bosque," "Arbol de Fuente."
Additional information: The African tulip tree grows very rapidly and develops soft, brittle wood. Its flower buds are full of water. Sometimes, children use them as water pistols. Pollinating birds use the leaves as an airstrip because they cannot drink nectar in flight. One seldom sees the yellow flowering species.

Rose of Venezuela

Brownea grandiceps
Family: *Caesalpiniaceae*, Carob plants

Most important features: This is a small tree or high bush with pinnules. The red flower heads are almost spherical and up to 10 inches (25 cm) wide.
Growth structure: The plant grows up to 33 feet (10 m) high, either with a thin trunk and dense, umbrella-shaped crown or with a wide, bushy crown.
Leaves: The leaves are alternating, up to 12 inches (30 cm) long, and bipinnately divided. They have seven to eleven pairs of pinnules, each one acute. At first, young branches and their leaves are flabby and hanging. They are pale or reddish in color. Later, the leaves turn green and become erect, shedding its deciduous leaves (see pages 16–17).
Flowers: The flowers are pale to bright red. Each inflorescence has several dozen flowers. The stamens are slightly longer than the corollas. The anthers are yellow.
Fruits: These are bean-shaped and grow up to 10 inches (25 cm) long.
Occurrence: One finds these in parks and gardens. As the name suggests, they originated in Venezuela.
Other names: German "Rose von Venezuela"; Spanish "Rosa de Montaña."
Additional information: The size and color of the flowers would indicate that birds are the pollinators. However, more often one sees numerous bees as active pollen and nectar thieves. Most of the other twelve *Brownea* species are cultivated, especially the scarlet flame bean (*B. coccinea*) and the large-leaved *B. macrophylla*. Both have longer filaments. *B. coccinea* has smaller capitula with only about twenty to thirty flowers; *B. macrophylla*, on the other hand, has only three to six pairs of pinnules, up to 12 inches (30 cm) long, per leaf. Even more rare is the white flowering *B. leucantha*.

Rain Tree

Albizia saman
Family: *Mimosaceae*, Mimosa plants

Most important features: The top is very large, noticeably wider than it is high. The branches are almost always full of epiphytes (see pages 18–19). The inflorescences look like powder puffs, pink outside and white inside.
Growth structure: The rain tree grows up to 100 feet (30 m) tall, but the trunk itself is only 7 to 10 feet (2 to 3 m) tall. The branches are flat and rise at an angle. The top is up to 165 feet (50 m) in diameter, umbrella-shaped, and flatly vaulted on the upper side.
Leaves: The leaves are alternating and bipinnately divided. They have small, round, slightly crooked pinnules, which fold themselves together at night or in very cloudy weather.
Flowers: The individual flowers are small and connected to what look like powder puffs about 2 inches (5 cm) in diameter. The puffs are created by numerous filaments and by a few pistils. The remaining organs are insignificant.
Fruits: The fruits are oblong, up to 12 inches (30 cm) long, often slightly bent, and reminiscent of large beans.
Occurrence: Although the rain tree is indigenous to many places, originally, it was found from Central America to the Amazon.
Other names: English "Monkey Pod Tree," "Cow Tamarind"; German "Regenbaum"; Spanish and Portuguese "Samán."
Additional information: Below the rain tree, it sometimes seems to rain, even in bright sunny weather. This is the fault of cicada, which prick the tree and live off its sugary juice. The insects then excrete the water they do not use. Single drops are hardly recognizable, but if the insects are present in large numbers, a kind of rain develops in the shadow of the tree. Because the rain tree reaches imposing dimensions, it is often planted individually.

Asoka Tree

Saraca indica
Family: *Caesalpiniaceae*, Carob plants

Most important features: Leaves are pinnately divided, and most of the flowers are orange, but the youngest ones are usually yellow. The oldest ones are always darker with long filaments.
Growth structure: The asoka tree grows up to 78 feet (24 m) high with a very dense crown. On occasion, it is significantly smaller; sometimes it is only bushy.
Leaves: These are alternating. The midrib is 2.8 to 20 inches (7 to 50 cm) long with two to seven pairs of tapering leaves. These often have short petioles and leaves up to 8 inches (20 cm) long and 2.4 inches (6 cm) wide. On occasion, the leaves are up to 12 inches (30 cm) long and 4 inches (10 cm) wide; the lowest pair are smallest.
Flowers: The flowers are umbrella-shaped. The spherical panicles are often up to 4 inches (10 cm) in diameter. They may grow up to 8 inches (20 cm) in diameter. The individual flowers have long petioles with four brightly colored sepals up to .4 inch (1 cm) long and six to ten filaments up to 1.4 inches (3.5 cm) long. The ovary is on the edge of the narrow flower tube. The tube, including the pistil, is about as long as the stamens.
Fruits: The dark red fruits are flat, 2.4 to 10 inches (6 to 25 cm) long, .8 to 2.4 inches (2 to 6 cm) wide, and .4 inch (1 cm) thick. The two halves fall apart when the fruits are ripe. They have four to eight seeds up to 1.6 inch (4 cm) long.
Occurrence: The asoka, a popular ornamental tree, originally, grew from India to Southeast Asia.
Other names: English "Sorrowless Tree"; German "Asokabaum."
Additional information: The name "asoka" is used for this species and for the very similar *S. Asoca*; perhaps they are not different species at all. Buddha was supposed to have been born under an asoka tree, which explains the fact that these trees are often found on temple grounds. In Hinduism, asoka flowers, with their distinctive fragrance, play a role in sacrificial offerings. The red saraca (*S. declinata*), with inflorescences on stronger branches, and the yellow saraca (*S. thaipingensis*) are also planted occasionally.

Silky Oak

Grevillea robusta
Family: *Proteaceae*, Proteus plants

Most Important Feature: The leaves are fern-like and pinnately divided. The bottoms are covered with silver gray hair. The inflorescences are brushlike and yellow orange in color.
Growth structure: Silky oaks grow up to 100 feet (30 m) tall with a straight, continuous trunk and a loose, narrow, straight crown.
Leaves: The leaves are alternating. They are up to 10 inches (25 cm) long with eleven to twenty-three pinnas, which are deeply incised to pinnately divided.
Flowers: Yellow to orange, possibly even red, these grow in simple racemes up to 5.2 inches (13 cm) long (like a huge toothbrush). They often appear in groups on older branches. Individual flowers grow about .8 inch (2 cm) long on .8 inch (2 cm) long petioles. For a long time, the pistil is bent out in the shape of an arch from the perianth. Later, it becomes straight with a thickened stigma in the shape of a head.
Fruits: The fruits are boat-shaped with brown follicles. They are about .8 inch (2 cm) long with a long bill and one or two winged seeds.
Occurrence: Originally from Australia, silky oaks have been planted in many tropical and subtropical areas and are often overgrown.
Other names: German "Australische Silbereiche"; Spanish "Pino Australiano."
Additional information: The Australian silky oak is the most frequent representative of the Proteus plants in the tropics; however, it is not related to our oaks. Because it can withstand arid conditions, wind, and the blazing sun well, it is used to reforest stripped areas, as well as to protect more sensitive crops, such as coffee and tea. Often, it grows wild and establishes the tone of the landscape. In Australia, these large trees have become rare because they are popular for carpentry. Two other species which also belong to this family are the macadamia nut tree (*Macadamia integrifolia*), which has simple leaves and insignificant flowers, and the South African protea (*Protea spp.*), which, once in a while, can be found in flower shops.

Horsebean

Parkinsonia aculeata
Family: *Caesalpiniaceae*, Carob plants

Most important features: The branches have thorns arranged in pairs on the bottom of the leaf. The leaves have tiny, widespread pinnules. The flowers are yellow, but the highest petal of older flowers is brownish orange.
Growth structure: The horsebean grows up to 33 feet (10 m) high with a short trunk, greenish bark, and a wide, loose top. Most of the branches droop like a weeping willow's branches.
Leaves: The leaves alternate. Two to six leaves start from one point. Actually, the leaves are bipinnately divided with an extremely shortened midrib. Each leaf is up to 16 inches (40 cm) long with a slightly flattened midrib and forty to sixty tapering leaves, which are up to .3 inch (8 mm) long. The leaves become smaller and smaller toward the tapered end and are completely missing in the final part.
Flowers: The flowers grow in simple racemes up to 8 inches (20 cm) long and about 1 inch (2.5 cm) wide with five petals. The highest one has a petiole and shows reddish spots.
Fruits: Similar to beans, these are up to 6 inches (15 cm) long and .3 inch (8 mm) thick. They hang down and are dark brown and leathery with slight indents between the few seeds.
Occurrence: Originally, they grew in warm areas of the Western Hemisphere. Now they are cultivated in drier locations and are widely naturalized in southern Africa.
Other names: English "Jerusalem Thorn," "Mexican Palo Verde"; German "Jerusalemdorn"; Spanish "Espina de Jerusalem," "Espinillo," "Flor de Mayo."
Additional information: The horsebean is often planted as a hedge because it doesn't require much care and is able to withstand strong wind. In dry weather, it first sheds its lamella and later its midribs. This does not damage it in any way. The related palo verde (*P. microphylla*) can live for years without water.

Golden Shower

Cassia fistula
Family: *Caesalpiniaceae*, Carob plants

Most important features: The leaves are large and bipinnately divided. The flowers are yellow and grow in long, hanging simple racemes. The fruits are bar-shaped.
Growth structure: This tree seldom grows more than 33 feet (10 m) high. It has a short trunk with a wide and loose crown.
Leaves: The leaves are alternating and up to 20 inches (50 cm) long with four to eight pairs of pinnules with a length of up to 8 inches (20 cm). During the blooming period, many of the leaves are shed, but the tree is never entirely leafless.
Flowers: The flowers grow in simple racemes about 12 inches (30 cm) long. Occasionally, they grow up to 32 inches (80 cm) long. They are about 1.6 inches (4 cm) wide with five short petioles. The flowers have pale to bright yellow petals. The ovary and the lower three stamens are bent upward.
Fruits: These grow up to 24 inches (60 cm) long. They are bar-shaped and brown black in color with a roll of coinlike stacked seeds surrounded by a sticky, brown pulp.
Occurrence: One finds golden shower in all tropical areas, but originally, they were from India and Sri Lanka.
Other names: English "Indian Laburnum," "Pudding Pipe Tree," "Purging Cassia"; German "Indischer Goldregen," "Mannabaum," "Röhrenkassie"; Spanish "Lluvia de Oro."
Additional information: The fruits of the Indian golden shower are sold as manna on the market. The sweet fleshy fruit is a mild laxative. The tree is popular along streets, except for the large amount of fruit which falls. For this reason, the rain tree (*Cassia x nealae*) is often preferred. Its flowers show all shades between yellow and pink; however, as a hybrid (along with the pink flowering *C. javanica*), it is sterile and produces no fruit. Numerous other species are also cultivated; most of them have yellow flowers in smaller panicles at the ends of the branches (see page 128).

Copperpod Tree

Peltophorum pterocarpum
Family: *Caesalpiniaceae*, Carob plants

Most important features: The leaves are bipinnately divided. The inflorescences and buds have dense red hair. The flowers are yellow. The fruits are flat and reddish brown.
Growth structure: The copperpod tree grows up to 115 feet (35 m) high, but only about 33 feet (10 m) when cultivated. The crown is wide and shaped like an umbrella.
Leaves: These are alternating and up to 24 inches (60 cm) long. They have four to fourteen pairs of pinnately divided leaves of the first order, which have about twenty to forty tapering leaflets with one to two long small leaves with spherical or scalloped tips.
Flowers: The flowers grow in twenty to forty long panicles up to 1.8 inch (4.5 cm) long with five sepals and five seemingly wilted petals that are hairy at the bottom. The flowers also have ten stamens and an ovary with a short petiole.
Fruits: The fruits are 5.6 inches (14 cm) long and 1 inch (2.5 cm) wide. They are pointed with parallel veins and are often slightly pulled in between the three to four oblong seeds.
Occurrence: The yellow flamboyant is often used as an ornamental tree and as a shade tree for plantations. Although it originated in Sri Lanka, it spread to Southeast Asia and Northern Australia.
Other names: English "Yellow Flame Tree," "Yellow Poinciana"; German "Gelber Flamboyant"; Spanish "San Francisco."
Additional information: The copperpod tree offers an attractive picture all year round. It flowers for months and produces new, bright copper red fruits, which stay on the tree even when it sheds its blossoms and leaves for a brief time during the dry period. On Java, people produce a dark brown batik color from its bark. It is also cultivated in South Africa. Its sister species, *P. africanum*, is called the weeping wattle because it is tapped by cicadas in the same way the rain tree is tapped (see page 64).

Sweet Thorn

Acacia karroo
Family: *Mimosaceae*, Mimosa plants

Most important features: The branches have white thorns up to 2.8 inches (7 cm) long. Leaves are bipinnately divided; inflorescences, spherical and yellow. The fruit is bean-shaped.
Growth structure: The sweet thorn grows up to 50 feet (15 m) high. The top of the tree is usually round, but occasionally it is bushy.
Leaves: The leaves are alternating and up to 7.2 inches (18 cm) long. A small gland sits on the petiole and midrib between the two to six pinnately divided leaves of the first order. Each carries five to twenty tapering leaves that are up to .3 inch (8 mm) long.
Flowers: These grow in dense inflorescences about .4 inch (1 cm) long. They look like powder puffs. Among the many individual flowers, only the filaments are recognizable.
Fruits: The fruits are approximately 6.4 inches (16 cm) long and .4 inch (1 cm) thick, slightly stringy, and bare. Between the seeds, they are lightly pulled in. The fruits are brown when ripe and split into two parts.
Occurrence: This is one of the most abundant trees in Africa. Outside of Africa, one is more likely to see other, though similar species.
Other name: German "Süßdorn."
Additional information: The sweet thorn is the most plentiful of about three hundred pinnately divided *Acacia* species (compare with page 102). Unlike many other *Acacia*, it is not poisonous and can be eaten by wild animals. However, when animals eat it, root shoots develop and spread in large numbers in over-grazed areas. The bark of the sweet thorn is used for tanning and in the production of robes. Aggressive ants live in the swollen thorns of *A. drepanolobium* and some Central American species (e.g. *A. sphaerocephala*, see photo at right). The ants feed on the nectar of the leaf glands and on special food bodies created at the tip of the pinnules. In return, the ants attack everything that comes too close to their tree.

Nitta Tree

Parkia speciosa
Family: *Mimosaceae*, Mimosa plants

Most important features: The leaves are bipinnately divided. The yellow flowers are pear-shaped and hang from long petioles (see photo at left). The green fruits are oblong. Several grow from a wooden stalk that is thick-ened toward the end.
Growth structure: The nitta tree grows up to 100 feet (30 m) tall. It has a sweeping top. The young branches often have soft hair.
Leaves: The alternating leaves grow up to 15 inches (36 cm) long. Each has fourteen to eigh-teen pairs of pinnately divided leaves of the first order that carry approximately sixty to seventy small tapering leaves which are .3 inch (8 mm) long and approximately .1 inch (2 mm) wide.
Flowers: The flower heads are 2.4 inches (6 cm) long and grow from long, branched petioles hanging from the top. The small individual flowers have tubular corollas.
Fruits: Often four to ten of these hang together from one petiole. The individual fruits are flat. They grow up to 18 inches (45 cm) long and 2 inches (5 cm) wide. Often, the fruits are twisted. They each have twelve to eighteen oval seeds that are up to 1 inch (2.5 cm) long.
Occurrence: Originally, the nitta tree was from Southeast Asia, and it is often cultivated there. One seldom sees this tree outside of Asia.
Other names: English "Pete Bean"; German "Petebohne"; Malayan "Petai" (also adopted into French and Spanish); Thai "Sato."
Additional information: Because bats often pollinate the flowers of the nitta tree, the flowers must be hanging in the air in order to be accessible. For this reason, the fruits on the tree catch one's eye immediately. The people of Southeast Asia eat the seeds of the fruit raw, cooked, or roasted, often in a spicy sauce. The taste is similar to garlic, and eating the sauce transfers a similar smell to the eater. The young leaves are also eaten as vegetables.

Chinaberry

Melia azedarach
Family: *Meliaceae*, Cedar plants

Most important features: The leaves are bipinnately divided. The numerous, small flowers are violet. The fruits are yellow and spherical.
Growth structure: The tree grows up to 50 feet (15 m) high. The loose top grows out of the firm branches.
Leaves: These are alternating and 8 to 16 inches (20 to 40 cm) long. The pinnately divided leaves grow up to 3.2 inches (8 cm) long with coarsely serrated edges.
Flowers: The flowers grow in loose panicles up to 12 inches (30 cm) long. The flowers are .8 inch (2 cm) wide with five pale lilac petals and one darker stamen tube that is purple to violet. Typically, the flowers unfold at the same time as the leaves.
Fruits: The fruits grow up to .6 inch (15 mm) long. They remain on the tree during the leafless season and become wrinkly around the pit.
Occurrence: Originally from Asia and Australia, the chinaberry has been naturalized in many tropical and moderate areas.
Other names: English "Indian Lilac," "Persian Lilac," "Pride of India," "Seringa," "White Cedar"; German "Paternosterbaum," "Chinesischer Holunder," "Indischer Zedrachbaum," "Paradiesbaum," "Perlenbaum," "Persischer Flieder"; French "Lilas des Indes"; Spanish "Alelí," "Arbol Paraíso," "Lilayo," "Jacinto," "Pasilla."
Additional information: The large number of folk names indicates that this a very popular species. Folk medicine uses almost all parts of the tree. In addition, the reddish wood is used in the production of furniture and musical instruments. Although the yellow fruits appear attractive, they are poisonous. The pits of the fruits are often used to create "pearls" for jewelry and rosaries. An extract of the leaves is used to repel locusts. The neem (*Azadirachta indica*), a related species, supplies cattle food, wood, seed oil, and the strongest insecticide of plant origin.

Carambola

Averrhoa carambola
Family: *Oxalidaceae*, Sour clover plants
Most important features: This is a small tree or high bush with pinnately divided leaves. The flowers range from pink to violet and grow in red, bunchlike panicles. The fruits are yellow and star-shaped in cross-section.
Growth structure: The carambola grows up to a height of 40 feet (12 m). Often, however, it is much smaller with a very short trunk.
Leaves: The leaves are alternating and grow up to 8 inches (20 cm) long with five to eleven pinnules. Those opposite are up to 3.2 inches (8 cm) long. The leaves are slightly larger at the tip than at the base.
Flowers: These are up to about 1 inch (2.5 cm) in diameter and grow in groups of five. The petals are pink white to violet. The inflorescences are much shorter than the leaves and grow at stronger branches.
Fruits: In outline, the fruits are ovate to elliptical and up to 6 inches (15 cm) long. They have four to six distinct longitudinal ribs. The flesh is firm and glassy and often seedless. The taste is sweet-sour and may be very strong.
Occurrence: The carambola grows in the tropics and in frost-free subtropics. It was originally from Southeast Asia.
Other names: English "Chinese Gooseberry"; German "Sternfruchtbaum"; Spanish "Tamarindo Chino."
Additional information: The fruit is often cut into slices and used as a decorative garnish. In Asia, the fruit is used in a variety of ways. Slightly salted, it is considered a great thirst quencher. It may be candied with sugar, cooked or uncooked, or eaten as a jelly. Its juice provides a sour touch to cocktails. Its flowers are used in salads. Bilimbi (*Averrhoa bilimbi*), a closely related species, is cultivated less frequently. Its leaves grow up to 24 inches (60 cm) long and have many more pinnules. The inflorescences often grow directly out of the trunk. It has dark red flowers and green, very sour fruits that are only slightly longitudinally grooved.

Pepper Tree

Schinus molle
Family: *Anacardiaceae*, Sumach plants

Most important features: The hanging branches resemble the weeping willow. The hanging leaves are pinnately divided and a light gray green in color.
Growth structure: The pepper tree grows up to 50 feet (15 m) high. Often the trunk is bent. When the trunk is injured, it oozes a resin with a wonderful smell.
Leaves: The leaves are alternating. They grow up to 8 inches (20 cm) long with eleven to forty-one pinnules. These are 1.2 to 2.8 inches (3 to 7 cm) long, slender, and leathery.
Flowers: The flowers are very small and pale yellow. They hang in groups of five on panicles that are up to 8 inches (20 cm) long. Male and female flowers grow on separate plants.
Fruits: These are pink to red and spherical. The flesh is thin and dries rapidly. The pits are about .2 inch (5 mm) thick.
Occurrence: One finds the pepper tree in areas with a hot dry period, in the Mediterranean area, and in South Africa. The pepper tree originated on the southern point of South America.
Other names: German "Peruanischer Pfefferbaum"; Spanish "Arbol de Pimienta," "Pirú."
Additional information: The fruit of the pepper tree is about the same size as the fruit of the real pepper (see page 268). The taste is strong. Historically, it was a very expensive spice. It is still found in various pepper crossbreedings. Christmasberry (*S. terebinthifolius*), a related species, is often cultivated. Its branches do not hang, and the pinnules are wider and dark green. Its bright red fruits are poisonous and remain on the tree for a long time. In some areas of the world, it is used as a Christmas ornament instead of the usual North American holly (*Ilex spp.*).

Drumstick Tree

Moringa oleifera
Family: *Moringaceae*, Ben plants

Most important features: The leaves are bipinnatedly to triple-pinnatedly divided. The flowers are cream white with four small, drooping petals and one larger, erect petal. The fruits are very oblong and slightly triangular.
Growth structure: The drumstick tree grows up to 33 feet (10 m) high with a light gray, smooth bark and a loose, irregular top.
Leaves: The leaves are alternating. They grow up to 24 inches (60 cm) long; the pinnately divided small leaves are only .4 to .8 inch (1 to 2 cm) long, elliptical, spherical, and tapering. The leaves are light green and shed easily.
Flowers: The flowers are .8 inch (2 cm) long on panicles up to 12 inches (30 cm) long. The stamens are slightly bent. Five have yellow anthers, and three to five have no anther. The ovary is on a short petiole.
Fruits: These hang. They are up to 24 inches (60 cm) long and 1 inch (2.5 cm) thick, with nine longitudinal ribs. Three of the ribs are stronger than the others. For a long time, the fruits are green. Then, they turn brown with three open flaps. The seeds have three thin wings.
Occurrence: Originally from Northwest India, the drumstick tree is cultivated in many tropical areas and is often overgrown.
Other names: English "Horseradish Tree"; German "Meerettichbaum"; French "Ben Ailée"; Spanish "Ben," "Maranga"; Portuguese "Muringueiro."
Additional information: The drumstick tree is named for its roots, which taste like horseradish. They are sometimes used in a similar manner. However, extracts from the leaves and the bark are also used to treat infections. Even its aroma may have an antibiotic effect. In Asia, young branches, leaves, and fruits are served as vegetables. Oil from the tree does not become rancid and has multiple uses, from cooking oil to soap and from creams to high-quality oil paints.

Indian Coral Tree

Erythrina variegata
Family: *Fabaceae*, Butterfly plants

Most important features: The branches are thorny. Each leaf has three large folioles and yellow patterns. The flowers are bright red and grow in dense simple racemes at the ends of the branches.
Growth structure: The tree grows up to 65 feet (20 m) high and often as wide.
Leaves: The leaves are alternating and are often shed in the dry period. They are widely rhombic to widely ovate and up to 8 inches (20 cm) long and wide. The leaves are tapered and without thorns. The young leaves are hairy on the bottom side.
Flowers: These grow in long panicles of up to 8 inches (20 cm), often on the almost leafless trunk. One large petal, up to 2.8 inches (7 cm) long, encompasses the ten filaments for a long time. The remaining four petals are much shorter.
Fruits: The fruits are sausage-shaped. They grow up to 16 inches (40 cm) long and 1.2 inch (3 cm) thick and are slightly pulled in, like a peanut, between the 1.2 inch (3 cm) large, reddish seeds.
Occurrence: One finds Indian coral trees in all tropical areas. The trees occur naturally from East Africa to Asia and up to the Pacific.
Other names: English "Tiger's Claw"; German "Indischer Korallenbaum."
Additional information: Numerous forms of the Indian coral tree are cultivated. For example, some have white flowers or a small top. However, the most popular is a species whose leaves have yellow along the main veins. Many of the remaining one hundred *Eyrthrina* species are cultivated (see also page 122). *E. abyssinica* (red hot poker tree) differs because of its bent thorns, which often appear on the bottom side of the leaf and its spherical tips. It is only about half as large as the Indian coral tree and has light red flowers with long, slender cusps. *E. caffra* has bare, tapering leaves and even smaller flowers. Unlike most other species, *E. fusca* has not red, but orange to pale brownish yellow leaves.

Red Silk Cotton Tree

Bombax ceiba
Family: *Bombacaceae*, Cotton tree plants

Most important features: The trunk has fleshy thorns. The leaves are palmately cleft. The bright red flowers are strikingly large.
Growth structure: The trees grow up to 130 feet (40 m) high with thick branches. Large specimens often have swollen trunks and firm buttress roots. In gardens, these trees often grow bushy.
Leaves: The trees shed their leaves during the dry period. The leaves have long petioles with five to seven palmately divided leaves. The leaf in the middle is the longest, up to 12 inches (30 cm).
Flowers: These may grow individually or a few may grow together. They are 2.8 to 6 inches (7 to 15 cm) long with a cup-shaped calyx and five shiny petals. The many stamens are often filled to the rim with nectar. Sometimes, the flowers appear before the deciduous leaves sprout.
Fruits: These are oblong capsules, 4 to 6.8 inches (10 to 17 cm) long, with five flaps that break open. The capsules are filled with whitish cotton. The seeds are approximately .3 inch (8 mm) long.
Occurrence: The red silk cotton tree grows in all tropical areas. Originally it grew in South India and Sri Lanka.
Other names: German "Indischer Seidenwollbaum," "Asiatischer Kapokbaum"; Spanish "Malabarico."
Additional information: The red silk cotton tree supplies almost half the total kapok used worldwide. The quality of its capsuled cotton is not as high as that of the real kapok tree (see page 84), but it is used more frequently. Every part of the plant is used, from the wood to the stamens, which the people of Thailand use to dye their dishes red. The roots are supposed to act as a diuretic; the bark is used to treat diarrhea; the resin is a styptic, etc. Several related species are also cultivated. Most of them have white flowers, strongly curled petals, and a bunch of stamens reminiscent of a shaving brush. *Pseudobombax ellipticum* from Mexico is called the shaving brush tree; its flowers are mostly pink.

Floss Silk Tree

Ceiba speciosa
Family: *Bombacaceae*, Cotton tree plants

Most important features: The tree has a wide top and a thick trunk. The leaves are palmately cleft. When the tree is leafless, it is covered with large, pink flowers. When the leaves sprout, the encapsulated cotton hangs down like cotton swabs.
Growth structure: Floss silk trees grow to be approximately 50 feet (15 m) high. The trunks of young trees are green and have pointed thorns; the trunks of older trees are gray, swollen, and slightly barrel-shaped.
Leaves: During the dry period, the leaves are shed. The long petioles with their five to seven leaves grow up to 4.8 inches (12 cm) long. They are palmately divided.
Flowers: The flowers grow in bunches on short, thick branches. The flowers are up to 6.4 inches (16 cm) in diameter. They have five petals. Most of the flower is pink to purple, but at the bottom, it is yellowish white, and in between, it is freckled. The stamens are connected to a midrib that is 3.2 inches (8 cm) long.
Fruits: These are pear-shaped and up to 8 inches (20 cm) long and 2 inches (5 cm) thick. The flaps of the capsules fall off so that the bright, white cotton hangs out.
Occurrence: These trees often grow in parks. Originally, they were from southern Brazil and Argentina.
Other names: German "Chorisie," "Brasilianischer Florettseidenbaum"; Spanish "Ceiba del Brasil."
Additional information: The floss silk tree grows in semidry forests and is often leafless for months. Twice a year, it offers a spectacular picture. In full bloom, the whole tree appears pink, and during the flowering season, the cotton swabs shimmer visibly in the sun until the wind rips them apart and carries them off. The flowers of the less frequently cultivated white floss silk tree (*C. insignis*) are yellowish with purple sprinkles. However, these rapidly fade. It has a thick trunk that holds water.

Pink Trumpet Tree

Tabebuia rosea
Family: *Bignoniaceae*, Trumpet tree plants

Most important features: The leaves are palmately cleft. The flowers have five somewhat wrinkled, pink corolla lobes and yellowish tubes.
Growth structure: The pink trumpet tree grows up to 80 feet (25 m) tall, but it is often smaller. The bark is cracked linearly.
Leaves: The five small leaves are elliptical to tapering. The one in the middle is up to 15 inches (35 cm) long and 7.2 inches (18 cm) wide. The stalk is 4.4 inches (11 cm) long. The ones on the sides are smaller and shorter.
Flowers: The flowers grow in panicles at the ends of the branches. The flowers are up to 4 inches (10 cm) long. The calyx has two parts. The corolla is almost white to red violet. Inside are two longer and two shorter stamens.
Fruits: These are bar-shaped capsules up to 16 inches (40 cm) long and .6 inch (1.5 cm) thick. They have two flaps that break open, releasing many winged seeds.
Occurrence: One of the most frequently cultivated ornamental trees, the pink trumpet tree originated in the Western Hemisphere.
Other names: English "Pink Poui," "Pink Tecoma"; German "Rosa Trompetenbaum," "Rosa Ipé-Baum"; French "Poirier Rouge"; Spanish "Amapa Rosa," "Apamate," "Mano de León," "Macuelizo," "Palo de Rosa," "Roble Colorado."
Additional information: The pink trumpet tree grows wild in areas of alternating humidity. There, it flowers after it has shed its leaves. When cultivated in areas that are continuously humid, it does not shed its leaves all at once and flowers much more fragrantly. It also serves as a shade tree for coffee and cacao plantations. It is harvested for its gray brown wood, which is heavy, very durable, and reminiscent of oak. It is frequently used to produce furniture. Other species have yellow flowers.

Octopus Tree

Schefflera actinophylla
Family: *Araliaceae*, Ivy plants

Most important features: The leaves are very large and palmately divided. The flowers and fruits appear above the leaves in clusters on short petioles. They grow in radial inflorescences that are 3 feet (1 m) long.
Growth structure: These seldom grow taller than 33 feet (10 m). However, in the wild, they are supposed to be able to reach up to 100 feet (30 m). They often have only little, erect branches. When cultivated, they are often bred as bushes.
Leaves: These are dark green and shimmering. They grow at the ends of the branches on long petioles, seven to eighteen to a petiole. They are palmately divided and hang from the petiole like a spoke. They are up to 12 inches (30 cm) long and 4 inches (10 cm) wide.
Flowers: The flowers are small, pink to red. They grow in groups of three to nine.
Fruits: These are small and purple to almost black with four to nine pits.
Occurrence: Frequently used as an ornamental plant, the octopus tree was originally from New Guinea and northern Australia.
Other names: English "Umbrella Tree"; German "Strahlenaralie," "Regenschirmbaum," "Tintenfischbaum"; Spanish "Arbol Paraguas," "Pulpo."
Additional information: Like many other of the approximately 650 *Schefflera* species, the octopus tree can sprout and grow on other trees, send aerial roots to the ground, and finally overgrow its host (compare with the banyan, see page 120). Young plants are very easy to grow, which makes them ideal as house plants. Many other members of the ivy family are used as decorative green plants. Some are important for economic reasons. For example, the core of the rice paper plant (*Tetrapanax papyrifer*) is used to produce Chinese rice paper, and *Panax ginseng, P. pseudoginseng, P. quinquefolius*, and *Eleutherococcus senticosus* are used to produce ginseng.

Guiana Chestnut

Pachira aquatica
Family: *Bombacaceae*, Cotton tree plants

Most important features: The leaves are palmately divided. The flowers are very large and light with five long, slender petals and a brush of many stamens, which are red at the ends. The fruits are large and light brown with short hair.
Growth structure: Guiana chestnuts grow up to 65 feet (20 m) high with smooth, gray brown bark. Often the top is small; sometimes the trees grow buttress roots.
Leaves: The leaves are alternating and up to 12 inches (30 cm) long. Five to nine leaflets grow on a short petioles.
Flowers: The flowers grow individually or up to three in a bunch. Each flower is 8 to 15 inches (20 to 35 cm) long. The petals are up to .8 inch (2 cm) wide and cream white to greenish or bluish yellow. Occasionally, they are red at the pointed end. The more than two hundred filamentous stamens are white or yellow at the bottom and red at the top.
Fruits: The fruits are almost spherical to tapering or elliptical and 8 to 16 inches (20 to 40 cm) long. They have five flat linear furrows along which the capsules open when ripe. Inside are a few oblong, brown seeds up to 2.4 inches (6 cm) long.
Occurrence: The Guiana chestnut is often planted along streets as a decorative tree. Originally, it was from the Western Hemisphere.
Other names: English "Provision Tree," "Saba Nut"; German "Wasserkastanie"; French "Cacao Sauvage," "Châtaignier Marron"; Spanish "Amapola," "Castaño de Agua," "Oje," "Palo de Boya," "Tetón," "Zapote de Agua."
Additional information: The Guiana Chestnut grows wild in areas that regularly flood. Its seeds are able to float and often sprout in water. When they are washed ashore, the seeds continue to grow. The seeds, often also called saba nuts, are edible if cooked or roasted. The *P. insignis* with its red petals and white stamens is also frequently planted.

Buttercup Tree

Cochlospermum vitifolium
Family: *Bixaceae*, Annato plants

Most important features: The leaves are deeply, palmately lobed. The tree produces a yellow, milky sap. The large yellow leaves appear after the tree is leafless. The capsules have fuzzy, hairy seeds.

Growth structure: Although it can grow up to 65 feet (20 m) high and 20 inches (50 cm) thick, it is often smaller, more like a little, branched tree. The bark on young branches is red brown; on old branches, it is silvery.

Leaves: These are alternating with long petioles. The five to seven deeply lobed leaves have a heart-shaped base and are up to 16 inches (40 cm) long. The leaf edges are delicately serrated. The leaves drop during the dry period.

Flowers: The flowers grow in panicles at the ends of the branches. They are up to 4.8 inches (12 cm) wide with coarse sepals and five golden yellow petals. They are slightly lobed at the top and have numerous stamens.

Fruits: The fruits are ovate. They are capsules that are about 4 inches (10 cm) long with three to five flaps. When ripe, the outer part of the fruit cover separates itself from the inner. Inside are numerous seeds with fuzzy, white hair.

Occurrence: The butterfly tree is often found in areas with alternating humidity. Originally, the tree spread from Mexico to Peru to Brazil.

Other names: English "Wild Cotton"; German "Butterblumenbaum"; Spanish "Bototo," "Carnestolenda," "Poro Poro," "Rosa Amarilla."

Additional information: In the dry period, the "buttercups" continue shining from the bare trees. When cultivated, one often finds trees filled with rose-shaped flowers. Extracts from the bark are used for various medical purposes. The hairy seeds can be used as kapok (see page 84). In Asia, *C. religiosum* is more common. It, too, is called the buttercup tree. It has slightly thicker, knotty branches and smaller leaves that are lobed three to five times with a smooth edge.

Gold Tree

Tabebuia chrysantha
Family: *Bignoniaceae*, Trumpet tree plants

Most important features: Five palmately divided leaves grow opposite each other. Yellow flowers have five lightly wilted corolla lobes.

Growth structure: This tree grows up to 65 feet (20 m) tall. The bark is dark gray and scaled.

Leaves: The tree sheds its leaves during the dry period. The small, tapering leaves are hairy on the bottom. The middle leaf is up to 6.8 inches (17 cm) long and 3.6 inches (9 cm) wide on a petiole that is 1.2 inch (3 cm) long. The side leaves are smaller, and the petioles are shorter.

Flowers: The flowers grow in dense bunches at the end of branches, which are leafless toward the end of flowering season. The flowers are up to 2.6 inches (6.5 cm) long. The calyx has five lobes. The corolla tubes often have delicate, red stripes at the upper part of the inner petals. They are hairy inside and have two longer and two shorter stamens.

Fruits: Fruits are bar-shaped capsules up to 20 inches (50 cm) long and .8 inch (2 cm) thick. Two flaps break open to release many winged seeds.

Occurrence: The gold tree is one of the most popular of all the ornamental trees. Originally, it was from the Western Hemisphere.

Other names: English "Yellow Poui"; German "Goldbaum," "Gelber Ipé-Baum"; Spanish "Amapa Prieta," "Araguaney," "Cortez Amarillo," "Guayacán," "Matasilisguate," "Primavera," "Roble Amarillo," "Verdecillo."

Additional information: The gold tree is representative of a number of yellow flowering *Tabebuia* species (for pink flowering, see page 78). All carry the same folk name and are often cultivated as ornamentals. The hard, heavy wood is very popular. It grows in a variety of colors from yellow to dark olive brown with delicate yellow points. The wood of *T. guayacan* is considered the most durable in the Western Hemisphere. The trunks of this species can be found still standing in water where the forest was flooded during the Panama Canal construction.

Kapok Tree

Ceiba pentandra
Family: *Bombacaceae*, Cotton tree plants

Most important features: This is a massive tree with a thick trunk. The leaves are palmately divided. The flowers are whitish with five stamens. The fruits are filled with cotton.
Growth structure: The kapok tree grows up to 230 feet (70 m) high. The young trees have cone-shaped thorns growing at regular intervals along the trunk and the branches. These stand out almost straight, at close to right angles. Older trees have large buttress roots and umbrella-shaped tops.
Leaves: The tree sheds its leaves during the dry period. The long petioles have five to nine palmately divided leaves, each of which is 4 to 8 inches (10 to 20 cm) long.
Flowers: These grow in simple umbels with green, cup-shaped calyx and five reddish to greenish white petals. The flowers are about 1 inch (2.5 cm) long and have yellow anthers. The flowers often appear while the tree is leafless.
Fruits: These are oblong capsules up to 6 inches (15 cm) long with five flaps that break open. Each is densely filled with brownish white cotton and up to one hundred dark brown seeds.
Occurrence: Now found in all tropical areas, the kapok tree was originally from Central and South America.
Other names: English "Silk Cotton Tree"; German "Kapokbaum," "Wollbaum"; Portuguese "Sumaúma"; Spanish "Ceiba."
Additional information: In the Western Hemisphere, the kapok tree towers over the rain forest. Its flowers smell like sour milk and are pollinated by bats. The cotton, which grows inside of the fruits, is sold as silk cotton on the market (see also page 78). The fibers are too smooth to be woven; however, they are water-repellent and filled with air. They are used as filling for outdoor jackets and for life vests. The oil from the seeds has a high content of linoleic acid and is used as cooking oil.

Baobab

Adansonia digitata
Family: *Bombacaceae*, Cotton tree plants

Most important features: The trunk is short and strikingly thick. The few main branches often seem to extend downward. The leaves are palmately divided and are shed during the dry period. The strikingly large flowers are white and hang down.
Growth structure: The tree has a sweeping top. The trunk is often half as thick as it is tall. Sometimes, it is even thicker than it is tall, giving it a very strange shape.
Leaves: The leaves are alternating with often five to seven tapering, palmately divided leaflets, each up to 6 inches (15 cm) long.
Flowers: The flowers hang individually. They are white and grow up to 8 inches (20 cm) in diameter with five sepals and five fleshy petals. The petals are wavy and often rolled backward or downward. The numerous filaments are connected to a midcolumn about half of their length. After the flowering, this midcolumn and the petals fall off. The pistil is slightly longer than the filaments and curled backward or downward.
Fruits: The fruits are ovate or oblong, 4.8 to 8 inches (12 to 20 cm) long and gray in color. The peel is wooden, and the inside has a mealy, whitish flesh and numerous seeds.
Occurrence: Once in a while, one finds these trees planted in parks. Originally, they were native to the savannah belt of southern Africa (Namibia, Botswana, Zimbabwe, Mozambique, and South Africa).
Other name: German "Affenbrotbaum."
Additional information: The flowers emit an unpleasant smell at night and are pollinated by bats. The mighty trunk has very soft wood which can save large amounts of water, yet the tree is often bare for months. Old specimens reach 30 feet (9 m) in diameter, are often hollow, and sometimes even serve as a shack. Some of the numerous traditional uses of the baobab include magical elements, surely due to its bizarre shape.

Food Candle Tree

Parmentiera aculeata
Family: *Bignoniaceae*, Trumpet tree plants

Most important features: The branches are short and have thick thorns. The leaves grow in groups of threes and are palmately divided. They are wrinkly and greenish white and often have a reddish vein. The fruits are cucumber-shaped.
Growth structure: This tree grows to be about 33 feet (10 m) tall. It has a trunk and a wide top.
Leaves: The leaves are opposite or grow in bunches with widened petioles. They are elliptical in shape and .6 to 2.8 inches (1.5 to 7 cm) long and .2 to 1.2 inch (0.5 to 3 cm) wide. The middle leaf is larger than the ones on the side. The leaves are often serrated on the margin.
Flowers: Flowers grow individually or in bunches on the trunk and at the ends of the branches. The calyx is split at the bottom and 1 to 1.6 inch (2.5 to 4 cm) long. The corolla tube is 2 to 2.8 inches (5 to 7 cm) long and slightly bent with a four-lobed limb. The pistil and two of the four stamens project slightly from the tube.
Fruits: These are oblong and yellowish green. They are 4 to 10 inches (10 to 25 cm) long and 1.2 to 2 inches (3 to 5 cm) thick with clear longitudinal ribs and a bulge at the stipula of the petiole.
Occurrence: They are often found in the Western Hemisphere. Originally, from Central America, they have been cultivated and naturalized in Australia.
Other names: German "Kürbis-Kerzenbaum"; Spanish "Chote," "Cuajilote."
Additional information: The flowers exude a sweet fragrance at night and are pollinated by bats. Fruits are edible cooked or raw. The glassy flesh with its many flat seeds is reminiscent of cucumbers; however, the taste is similar to peas. In the Western Hemisphere, one can still find examples of the real food candle tree (*P. cereifera*). It has no thorns. Its fruit is 12 to 52 inches (30 to 130 cm) long but only 1 inch (2.5 cm) thick. Because of its size and because its skin is smooth and waxy, it looks rather like a handmade candle.

Rubber Tree

Hevea brasiliensis
Family: *Euphorbiaceae*, Wolf's milk plants

Most important features: The bark is smooth and light gray and often has long diagonal incisions. The tree produces an abundance of milky sap. The leaves are tripalmately divided. The large seeds are brown.
Growth structure: The rubber tree grows up to 100 feet (30 m) high. It is richly branched and often has a slender top.
Leaves: The leaves are alternating on long petioles. The tree sheds during the dry period. The palmately divided leaves are oblong. Usually they are about 6 inches (15 cm) long. On the bottom, they appear bluish and waxy.
Flowers: These are greenish white and about .2 inch (5 mm) long. They grow in panicles at the ends of the branches.
Fruits: The capsules grow up to 2 inches (5 cm) long. They open with an audible, loud bang, flinging the seeds—up to 1.4 inch (3.5 cm) long—50 feet (15 m) from the tree.
Occurrence: Rubber trees grow in all humid tropical areas, but primarily on plantations. The trees originated in the southern area of the Amazon.
Other names: German "Kautschukbaum"; French "Caoutchouc"; Portuguese "Seringa"; Spanish "Caucho."
Additional information: The rubber tree is the most important, technically useful plan of the tropics. It supplies about ninety-five percent of the world's natural rubber. The number of products made from this tree is estimated at 55,000. One obtains the milky sap, rich in latex, by cutting the bark diagonally. Crude rubber is sticky when hot and brittle when cold. The invention of the vulcanizing process (1839) led to a rubber boom in Brazil. In 1876, Henry Wickham brought 70,000 seeds to England. Approximately 2,400 of them thrived in the orchid houses of Kew Gardens. Most of the seedlings were shipped to the British colonies, where almost all of them died. Only twenty-two of the original seeds survived in the botanic garden of Singapore. Descendants of these seeds cover millions of acres (hectares), especially in Malaysia and Indonesia.

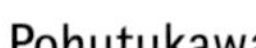

Pohutukawa

Metrosideros excelsa
Family: *Myrtaceae*, Myrtle plants

Most important features: The leaves grow opposite. Dense, white gray hair grows from the branches, the bottom of the leaves, and the calyxes. The inflorescences look like crimson red powder puffs.
Growth structure: The trees grow up to 65 feet (20 m) high. The top is wide and shaped from the bent, interwoven branches. When cultivated, it is often grown as a bush.
Leaves: The leaves are elliptical to tapering and up to 4 inches (10 cm) long. They are dark green, leathery, and rather dull. Often, the margin is curled. The leaves point toward the ground.
Flowers: The flowers grow in dense bunches with five fluffy, hairy sepals, five tiny, red petals, and a large number of filaments. The flowers are up to 1.6 inch (4 cm) long. They are crimson with yellow anthers and a yellow green nectar ring around the pistils.
Fruits: The capsules have three flaps and are up to .4 inch (1 cm) long. They are crowned by sepals.
Occurrence: One sees pohutukawa as an ornamental bush in gardens. It originated in New Zealand.
Other names: English "Christmas Tree"; German "Pohutukawa," "Eisenholz."
Additional information: The pohutukawa is one of five tree-shaped *Metrosideros* species of New Zealand. The name, Christmas tree, refers to the fact that it usually flowers at Christmas. Pohutukawa is frequently confused with *M. robusta;* however, *M. robusta* is rarely planted in gardens because it grows too rapidly. It differs from pohutukawa in its plain, often smaller leaves. The tree most often seen in Hawaii is *M. polymorpha.* Its name means "variform," reflecting the fact that its growth structure, leaf shape, and hairs are extraordinarily variable.

Malay Apple

Syzygium malaccense
Family: *Myrtaceae*, Myrtle plants

Most important features: The leaves grow opposite each other. The flowers are pink to red violet with four sepals and petals and hundreds of stamens. The fruits are pink white to red and fleshy. The remainders of the four sepals appear at the end.
Growth structure: The tree grows up to 80 feet (25 m) high with a very dense, often cone-shaped top.
Leaves: These are approximately 15 inches (35 cm) long and elliptical to tapering. They are leathery and dark green on the upper side and lighter on the bottom side.
Flowers: The flowers grow in groups of up to twelve, predominantly on older branches. They are hidden under deciduous leaves and reminiscent of a shaving brush. They are up to 2 inches (5 cm) wide. The petals are .8 inch (2 cm) long; however, they are insignificant compared to the stamens.
Fruits: The fruits are shaped like eggs or pears and 2 to 4.8 inches (5 to 12 cm) long with white flesh and one large seed.
Occurrence: They are usually found in gardens. As the name indicates, they are from Malaysia.
Other names: English "Mountain Apple," "Ohia," "Otaheite Apple," "Pomerac"; German "Malayenapfel," "Apfeljambuse"; French "Jamelac," "Pomme de Tahiti," "Pomme de Malaisie," "Pomme Malac"; Portuguese "Jambo"; Spanish "Manzana"; Malaya," "Marañon Japonés," "Pera de Agua," "Pomagás," "Pomalaca," "Pomarosa de Malaca."
Additional information: The Malay apple is cultivated as an ornamental and fruit tree. Its fruits taste something like a sweet apple. One often adds spices when cooking these fruits as desserts or jams. This is also true for the fruits of the other species. *S. samarangense* and the very similar water apple (*S. aqueum*) are waxy and vary in color from white to pink. They are slightly smaller with a spinning-top shape. However, the fruits of *S. Cumini* are plum-shaped and purple.

Pride of India

Lagerstroemia speciosa
Family: *Lythraceae*, Willow herbs

Most important features: The leaves grow opposite each other. The flowers often have six petals, are light violet to pink, and have a slightly wilted appearance.
Growth structure: Although this tree grows up to 100 feet (30 m) tall, when cultivated, it only reaches 30 to 40 feet (9 to 12 m). It has a very wide top.
Leaves: The leaves are tapering and up to 10 inches (25 cm) long. The tree sheds its leaves in the dry period. Shortly before dropping, the leaves take on a dark red color.
Flowers: The flowers grow in erect inflorescences at the ends of the branches. They are up to 15 inches (35 cm) long and up to 3.2 inches (8 cm) wide. The petals are usually pink (occasionally white), and the many stamens are yellow.
Fruits: These are spherical to ovate and 1.2 to 2.4 inches (3 to 6 cm) long. The fruits are brown with six flaps.
Occurrence: One frequently sees these growing on streets and in parks. They spread from India to China and northern Australia and the tropics in the Western Hemisphere.
Other names: English "Crepe Myrtle," "Queen Flower," "Rose of India"; German "Rose von Indien"; Spanish "Astromelia," "Astromero," "Flor de la Reina."
Additional information: The pride of India is popular as an ornamental tree, and its reddish wood is in high demand. Because it is durable even in tropical climates, builders prefer it for houses and boats. Of the fifty *Lagerstroemia* species, grape myrtle (*L.indica*) is also often cultivated. It is from East Asia. It remains smaller than the pride of India, and its leaves are only 2.4 inches (6 cm) long. They are more spherical, and the flowers are smaller. The petals appear even more distinctly wrinkled. Typically, the flowers are the same color as the pride of India; however, they can also be light red. Because it can tolerate winters of up to 14° F (–10° C), but needs hot summers, it also thrives well in some areas of the United States of America.

Autograph Tree

Clusia major
Family: *Clusiaceae*, Clusia family

Most important features: Leaves are large, thick, leathery, and narrower at the base. The flowers are large, pink to white, and hermaphroditic.
Growth structure: The autograph tree grows up to 65 feet (20 m) high. The bark of the trunk is smooth. When cultivated, it is often bushy with sticky resin and aerial roots.
Leaves: The leaves grow opposite each other and are aproximately 8 inches (20 cm) long and 4.4 inches (11 cm) wide. At the tip, the leaves are widely spherical and very stiff. They have a firm midrib and insignificant lateral veins.
Flowers: Flowers occur individually or paired and are 2 to 3.2 inches (5 to 8 cm) long with four to six wide sepals and four to eight petals. Male flowers have many stamens; females have six to sixteen stigmas on the spherical ovary.
Fruits: Fruits are almost spherical and 1.6 to 2.4 inches (4 to 6 cm) wide. They are streaked with green or red and retain red brown sepals. When ripe, they are star-shaped with six to sixteen flaps. The seeds are white with a red hull.
Occurrence: The autograph tree is native to Central America, the Caribbean, and the northern portion of South America. It is rarely cultivated in other parts of the world.
Other names: English "Scottish Attorney"; German "Klusie"; Spanish "Copey," "Tampaco."
Additional information: The tough leaves of the autograph tree remain on the tree for years even if they are injured. The name dates back to this characteristic; one can often see leaves with names carved into them. Like many figs (see page 120), the autograph tree can be epiphytic for part of its life. For this reason, it is sometimes referred to as a tree killer. Of the 150 *Clusia* species, several are occasionally planted. *C. grandiflora* has larger blooms but not as many of them. *C. minor*, on the other hand, has smaller blooms, which are more plentiful. The resin of all species can be used to seal boats.

Teak

Tectona grandis
Family: *Lamiaceae*, Labiate flowering

Most important features: The young branches are square-shaped. The leaves are large and grow opposite each other. On the top, the leaves are rough; on the bottom, they are hairy. The flowers are small and grow in richly branched clusters.
Growth structure: The trees grow up to 145 feet (45 m) tall. The bark of the slender trunks is gray and cracked. The top of the tree is slightly cone-shaped.
Leaves: The leaves are elliptical to ovate and slightly pointed. The veins protrude on the bottom side. The leaves are up to 16 inches (40 cm) long and 10 inches (25 cm) wide. The leaves on young trees can grow up to 32 inches (80 cm) long and 16 inches (40 cm) wide. The tree sheds its leaves during the dry period.
Flowers: These are approximately .4 inch (1 cm) wide. They grow in clusters of five to seven flowers. The clusters are up to 20 inches (50 cm) long. The flowers are all cream colored.
Fruits: The fruits are up to .6 inch (1.5 cm) wide. They are fleshy and have up to four pits. The fruits are covered by a dry, roundish hull that is about 1 inch (2.5 cm) wide. The hulls are yellow green to beige brown color (see photo at right).
Occurrence: These trees are native to the area from India to Thailand. They grew in forest areas with a distinct dry period. Today, one finds them in all tropical areas, especially on plantations.
Other names: English "Indian Oak"; German "Teakbaum"; French "Teck"; Indonesian "Djati"; Spanish "Teca."
Additional information: The teak produces one of the most precious of all tropical woods. It resists seawater, termites, and fungus. Unlike other, similarly resistant forms of wood, it is relatively easy to work with and attractively colored. Up to the middle of the nineteenth century, great amounts of teak were used in the construction of ships. Even today, it is still in great demand for furniture and veneers. In the past, the tree was mistakenly considered part of the Verbain family (*Verbenaceae*), but that error has been corrected.

Rose Apple

Syzgium jambos
Family: *Myrtaceae*, Myrtle plants

Most important features: The leaves grow opposite each other. The flowers are cream white. Each has four sepals, four petals, and hundreds of stamens. The fruits are fleshy and crowned by four greenish sepals.
Growth structure: The rose apple tree grows up to 40 feet (12 m) high. It has a wide, dense top. When cultivated, it is often kept as a bush.
Leaves: The leathery leaves are up to 8 inches (20 cm) long. They are oblong and delicately tapering. The upper side of the leaves is dark green and shimmering; the bottom side is lighter and duller.
Flowers: These often grow in groups of three to five, reminiscent of a shaving brush. The flowers are up to 2.4 inches (6 cm) wide. The petals are up to .8 inch (2 cm) long, but they are insignificant compared to the stamens.
Fruits: The fruits are round to pear-shaped and approximately 1.2 to 2 inches (3 to 5 cm) long. They are green to pale yellowish with one or two large, round seeds.
Occurrence: Originally from tropical Asia, they are often planted in gardens.
Other names: English "Plum Rose"; German "Rosenapfel," "Aprikosenjambuse"; French "Jambosier," "Pomme Rose"; Portuguese "Jambeiro," "Jambo Amarelo"; Spanish "Jambo," "Manzanito," "Pomarosa."
Additional information: The firm, pale yellowish flesh of the rose apple smells like a rose and has a sweet-sour taste. The fruits are eaten uncooked or made into jam. Because they only last for a short time, they are not usually exported. Several closely related species are also used for food (see Malay apple, page 88). More economically important is the clove (*S. aromaticum*). Its buds serve as a spice, and the oil distilled from it is used in toothpaste, as well as in the production of vanilla flavoring.

Australian Flame Tree

Brachychiton acerifolius
Family: *Sterculiaceae*, Cacao plants

Most important features: Often one tree will have both lobed leaves and simple leaves. The flowers are coral red and grow in many large inflorescences.
Growth structure: The tree grows up to 115 feet (35 m) high. The trunk is continuous to the top. The bark is smooth or slightly split lengthwise.
Leaves: The leaves are alternating on long petioles. The leaves are a shimmering dark green and up to 8 inches (20 cm) long. On younger trees, the leaves have three to five (even up to seven) long lobes. The one in the middle is the largest and may widen slightly above its base. On older trees, the leaves become more and more simple. Finally, they are ovate with three veins running from the base. Although the leaves drop during the flowering season, often this only occurs on the flowering branches.
Flowers: These are up to 1.2 inch (3 cm) long and grow in clusters to 16 inches (40 cm) long. The calyx is bell-shaped, short, and has five lobes with no petals. The midrib has either fifteen yellow anthers or five yellow carpels.
Fruits: The fruits are thick and bean-shaped, 4 to 8 inches (10 to 20 cm) long. They grow in groups of up to five, hanging on long petioles. The seeds are bright yellow. Hairs in the seeds can irritate the skin.
Occurrence: The tree originated in Australia, but now it also grows in South Africa.
Other names: German "Flammenbaum."
Additional information: If one knocks the trunk of the large Australian flame tree, the trunk sounds hollow. The very light, soft wood (sometimes used as a substitute for balsa wood) is the reason for this hollow sound. The tree can withstand enough frost to be cultivated in the Mediterranean area. The remaining thirty *Brachychiton* species originated in dry forests. Some of them can grow into bottle trees. The kurrajong (*B. populneus*), has smaller flowers that are pale green on the outside and red on the inside. It is a very popular street tree in Australia.

Scarlet Cordia

Cordia sebestena
Family: *Boraginaceae*, Rough leaf plants

Most important features: The leaves are as rough as sandpaper and quite stiff. The flowers are bright orange red, and the petals have a wilted appearance.
Growth structure: The tree grows up to 33 feet (10 m) high; however, most specimens are smaller. Often, they are only bushy.
Leaves: The leaves are alternating and elliptical, heart-shaped, or ovate. They grow up to 8 inches (20 cm) long.
Flowers: The tree has flowers almost all year long. They grow in simple umbels at the ends of the branches. The flowers are about 1 inch (2.5 cm) across with a short green calyx tube, a long corolla tube, and an extended limb with five to seven petals. The small, yellowish stamens are barely longer than the tube.
Fruits: The fruits are ovate and whitish with a large pit. They grow up to 1.2 inch (3 cm) long. The flesh is very sweet and slimy. The calyx is slightly fleshy.
Occurrence: Often an ornamental plant, the scarlet cordia was originally from the Caribbean to Florida and Venezuela.
Other names: English "Geiger Tree"; German "Sebestene," "Scharlachkordie"; Spanish "Vomitel," "No-me-olvide."
Additional information: *Cordia* is represented by about 320 species found throughout the tropics. Most of them, however, have white, rather insignificant flowers. Because of their very rough leaves, they are also known as sandpaper trees. *C. subcordata*, which the Polynesians brought to Hawaii, is very similar to the scarlet cordia; but it has shimmering leaves and lighter orange flowers. Its wood is valuable for carving. The fruits of the scarlet cordia are sucked like candy to treat coughs. The ancient Egyptians used the slightly larger sebesten plum (*C. myxa*) for this purpose.

Sandbox Tree

Hura crepitans
Family: *Euphorbiaceae*, Wolf's milk plants

Most important features: All parts of the trees have a milky sap. The trunk has cone-shaped thorns. The leaves are heart-shaped to ovate with firm lateral veins. The flowers are purple and strangely shaped. The green fruit is shaped like a miniature pumpkin or like the slices of an orange with ordered segments.
Growth structures: The tree grows up to 80 feet (25 m) with a short trunk and wide top.
Leaves: The leaves are alternating and grow on petioles that are 8 inches (20 cm) long. The leaf blades are up to 10 inches (25 cm) long, light green, pointed, and dentate.
Flowers: The male flowers (see photo at left) are button-shaped and about .2 inch (5 mm) long. Approximately sixty to eighty of them hang on a cone-shaped spike that is up to 2 inches (5 cm) long. The petiole is up to 4 inches (10 cm) long. The female flowers grow individually. They usually consist of a bottle-shaped ovary that is up to 2.4 inches (6 cm) long with fourteen to twenty ray flowers.
Fruits: The fruits are brown, flattened spheres up to 2 inches (5 cm) long and 3.2 inches (8 cm) wide. Each fruit consists of fourteen to twenty parts. The fruits explode, scattering the flattened seeds, which are up to .8 inch (2 cm) long, over a wide area.
Occurrence: They were originally from the Western Hemisphere where they were planted as street trees or as living fence posts. Today, they are frequently found in Asia.
Other names: German "Sandbüchsenbaum"; French "Sablier," "Arbre au Diable"; Spanish "Jabillo."
Additional information: The sandbox tree owes its name to the fact that people used its fruits to absorb ink before the invention of blotting paper. Like many similar plants, it contains a poisonous milky sap which, when touched, causes skin irritations and can even lead to blindness. The juice is used for fishing; simply dropping it into standing bodies of water attracts fish.

Cannonball Tree

Couroupita guianensis
Family: *Lecythidaceae*, Brazil nut plants

Most important features: The trunk has many, short leafless lateral shoots. The large, pink red flowers and fruits reminiscent of cannonballs grow from these shoots.
Growth structure: The tree grows up to 115 feet (35 m) high with a rounded top.
Leaves: The leaves are alternating and up to 12 inches (30 cm) long. They are tapering and elliptical to obovate in shape. They accumulate at the ends of the branches. The tree may shed its leaves several times a year.
Flowers: The flowers are present almost all year round. They only appear on the bare branches below the top. They are 2.4 to 4 inches (6 to 10 cm) long with six deep red to pink and yellowish petals. In the middle, hundreds of stamens grow together. These may be white, pink, yellow, or a combination. At the upper and lower sides, the color is clearly different.
Fruits: These are spherical, brown, and 6 to 9.6 inches (15 to 24 cm) in diameter. The peel is hard. The flesh is mushy with an unpleasant smell. The fruits contain many seeds.
Occurrence: The cannonball tree is usually found in parks. Originally, it was from northern South America.
Other names: German "Kanonenkugelbaum"; French "Arbre à Bombes"; Portuguese "Castanha de Macaco"; Spanish "Bala de Cañón," "Coco del Mono," "Taparon."
Additional information: The flowers of the cannonball tree spread an intense sweet smell at night. This attracts bats to pollinate the flowers. In South America, the tree almost always has both flowers and fruits at the same time. In Asia, on the other hand, it seldom produces fruits, perhaps because the right pollinators are missing. Wild boars break open the hard shell of the fruits and eat the flesh.

Orchid Tree

Bauhinia variegata
Family: *Caesalpiniaceae*, Carob plants

Most important features: The leaves are spherical with two lobes. The flowers are large, and the highest petal is darker than the other four.
Growth structure: The orchid tree grows up to 40 feet (12 m) high, but it is often smaller with long, slightly drooping branches. Occasionally, it grows as a bush.
Leaves: The leaves are alternating and grow up to 2 to 8 inches (5 to 20 cm) long. At the bottom, they are spherical to heart-shaped; at the tip, they are sinuated up to one-third of the length.
Flowers: The flowers are individual, or a few appear on panicles at the ends of the branches. The flowers are 3.2 to 4.8 inches (8 to 12 cm) long. They are often pink or pale-violet and have five stamens bent upward.
Fruits: These are up to 12 inches (30 cm) long and .8 inch (2 cm) wide. They are flattened, woody, smooth, and brown. When they are ripe, they burst into two parts.
Occurrence: Frequently found in gardens, the orchid tree originally spread from India to South China.
Other names: English "Buddhist Bauhinia," "Bull Hoof"; German "Orchideenbaum"; French "Deux Jumelles"; Spanish "Arbol Orquidea," "Calzoncillo," "Pata de Cabra," "Pata de Vaca," "Urape Morado," "Urape Orquidea."
Additional information: The flowers of the orchid tree are reminiscent of large orchids (see *Cattleya*, page 246). However, the two are not related. The flowers of the orchid tree can be light red violet, delicately pink or white, with pink stripes or a violet spot on the upper petal, or have yellow green veins. Many other of the approximately three hundred *Bauhinia* species are cultivated (see pages 172 and 202). In Asia, *B. blakeana* (see photo at right) is especially popular. It never produces fruits. Supposedly, it is a crossbreed between *B. variegata* and the very similar *B. purpurea*, which usually has slightly darker flowers with lighter veins. *B. monandra* has red spotted petals and only one stamen.

Hydrangea Tree

Dombeya wallichii
Family: *Sterculiaceae*, Cacao plants

Most important features: The large, heart-shaped to weakly lobed leaves often have seven main reticular veins. These run like rays from the starting point of the petiole. The flowers are pink red and hang from long petioles. **Growth structure:** The hydrangea is a small tree, rarely reaching 33 feet (10 m). Occasionally, it is grown as a bush.
Leaves: The leaves are alternating. The petioles are up to 10 inches (25 cm) long with large stipules. The leaf blades are up to 12 inches (30 cm) wide. On the upper side, they are rough; on the bottom side, they are smooth and hairy. The edge of the leaf is dentate.
Flowers: These are dense, semispherical, head-shaped simple umbels of up to 6 inches (15 cm) in diameter. At first, they are enclosed in large green whorls of bracts. The individual flowers are up to 1 inch (2.5 cm) long. Each flower has numerous stamens connected to one column; the anthers are yellow.
Fruits: The fruits are small capsules with many seeds. Almost until they are ripe, the fruits are enclosed by the wilted remains of the flowers.
Occurrence: One finds the hydrangea tree in parks and gardens. Originally, it was from Madagascar.
Other names: English "African Mallow," "Mexican Rose," "Pink Ball Tree"; German "Hortensienbaum."
Additional information: Despite its name, the hydrangea tree is not related to the *Hydrangea* but rather to the cacao. Of the approximately 230 *Dombeya* species, 190 originated in Madagascar. The remaining species are from Africa and from the Mascarene islands. Most of them have smaller inflorescences without whorls of bracts, and only a few are as attractive as the hydrangea tree. These are also seen less often. Some species, especially those in Africa, can endure a light frost, so they can also be cultivated in the Mediterranean area.

Calabash Tree

Crescentia cujete
Family: *Bignoniaceae*, Trumpet tree plants

Most important features: This is a small, rather erect tree. The leaves grow in bunches. The flowers have bent tubes and five wrinkly tips. The fruits are very large.
Growth structure: The calabash tree grows up to 33 feet (10 m) high with branches starting at the bottom. The branches are long and hang down. The younger ones are four-edged.
Leaves: The leaves are alternating. They grow in bunches on short stipules. The leaves are tapering to obovate and up to 4.8 inches (12 cm) long.
Flowers: The flowers grow on short petioles on the trunk and branches. They are up to 3 inches (7.5 cm) long. The green calyx has two lobes. The corolla is a wide tube that is bent toward the ground. Often, it is brown red.
Fruits: The fruits are spherical to elliptical. They are 4 to 16 inches (10 to 40 cm) in diameter. The fruits are green to yellow, but after the shedding, they are brown. The peel is woody and about .2 inch (5 mm) thick. The flesh is mushy and whitish to lightly bluish. Inside are many flat, black seeds.
Occurrence: One finds the calabash tree in parks and gardens, especially in areas with alternating humidity. Originally, it was from the Caribbean.
Other names: English "Gourd Tree"; German "Kalebassenbaum"; French "Calebassier"; Spanish "Calabasa," "Cujete," "Higüero," "Taparo," "Totumo."
Additional information: The woody peel of the fruits of the Calabash tree is used in the production of rhythmic music instruments and containers of many kinds. Often, the fruits are tied as they grow in order to produce the desired shapes. Some people of the Caribbean area eat the fruit, flesh, and seeds; however, one has to be cautious, as these are poisonous when uncooked. In Asia, one finds the closely related *Crescentia alata*. It differs from the calabash tree in its tripalmately divided leaves and wide petioles. Bats pollinate both species.

Calabash Nutmeg

Monodora myristica
Family: *Annonaceae*, Scale apple plants

Most important features: The waxy leaves are often slightly gray blue; the youngest ones are reddish. The flowers are big and grow in groups of three with bent edges that curl under. The outer petals are pale yellow with dark red spots.
Growth structure: The tree grows up to 115 feet (35 m) high with a dense top. It is often cultivated as a bush.
Leaves: The leaves are alternating and tapering, growing up to 20 inches (50 cm) long and 8 inches (20 cm) wide. At the bottom, they are spherical to heart-shaped.
Flowers: The flowers are up to 10 inches (25 cm) long. They hang individually from petioles. The flowers have three sepals and six petals. The three outer ones are up to 4 inches (10 cm) long, firmly bent and wavy; the three inner ones are up to 2 inches (5 cm) long and paler. They hang together. Inside are many small stamens and one carpel.
Fruits: These are almost spherical and up to 6 inches (15 cm) in diameter. The fruits are green to black brown with a woody peel and white flesh. Each contains seeds .4 to .8 inch (1 to 2 cm) long.
Occurrence: Originally from tropical Africa, one finds the calabash nutmeg in Africa and in the Caribbean, but rarely elsewhere.
Other names: English "African (false) Nutmeg," "Jamaica Nutmeg"; German "Kalebassen-Muskatnuss"; in Central Africa "Mbende."
Additional information: Several *Monodora* species are sold as African nutmeg. The fragrance of the seeds is similar to real nutmeg (*Myristica fragrans*), but the two are not closely related. Because of its bizarre flowers, these trees are known as orchid trees. Beside the calabash nutmeg tree, one also finds the closely related *Monodora tenuifolia* planted. It has delicate flowers. The outer and small inner petals are stretched. The latter touch each other with the tips.

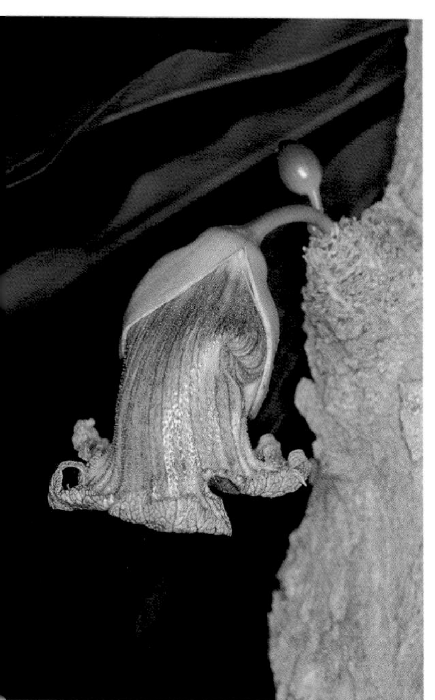

Ear-Pod Wattle

Acacia auriculaeformis
Family: *Mimosaceae*, Mimosa plants

Most important features: The leaves are slender and tapering. Often, they are slightly crescent-shaped and bent with three delicate, strong veins. The inflorescences look like yellow pipe cleaners. The fruits are bent and wavy.
Growth structure: This tree grows up to 80 feet (25 m) high. It is richly branched and seldom grown as a bush.
Leaves: The leaves are alternating and grow up to 6.4 inches (16 cm) long and 1.2 inch (3 cm) wide. The upper and lower sides are very similar.
Flowers: Many flowers grow together on spikes up to 3.2 inches (8 cm) long. These are often located in pairs in the bracts. The individual flowers are up to .2 inch (5 mm) long. The calyx and corolla are tiny; only the many yellow anthers are striking.
Fruits: These are flat and woody. When they are ripe, they are brown, and the two-part hull separates. The fruits are up to 2.8 inches (7 cm) long and .6 inch (1.5 cm) wide. They frequently grow in tangled groups.
Occurrence: Originally from New Guinea and northern Australia, this tree now grows in tropical Asia and Africa.
Other names: English "Northern Black Wattle"; German "Öhrchen-Akazie."
Additional information: Because the tree grows very rapidly and supplies a hard, dark red wood, it is often grown on plantations. Since the massive fires in parts of Southeast Asia in 1997 and 1998, it has aggressively edged out other vegetation in that area. Of the 1,200 species of the *Acacia* genus, about one-fourth have pinnas (see also page 70); the remaining ones have quite simple, more slender leaves. In young plants, these "leaves" are actually flattened and widened petioles as well as midribs of former pinnas. The inflorescences of most *Acacia* are yellow, spherical, and .4 to .8 inch (1 to 2 cm) long; in the inflorescences, one finds white flowers reminiscent of bottle brushes. Pink flowers occur much less frequently.

Cork Tree

Thespesia populnea
Family: *Malvaceae*, Mallow plants

Most important features: The leaves are heart-shaped. The large flowers are yellow to orange with a dark purple spot at the bottom of each petal. Only small, slender leaves grow on the outside of the calyx.
Growth structure: The cork tree grows up to 30 feet (9 m) tall with a short, often crooked trunk and a more spherical top. The young branches have tiny scales that are only visible under a magnifier.
Leaves: These are alternating and up to 6 inches (15 cm) long. They are shaped like arrows. The base is heart-shaped and has five light main veins. The upper side is slightly shiny. The stipules grow up to .3 inch (8 mm) long.
Flowers: These are up to 3.2 inches (8 cm) long with a cup-shaped calyx and a midcolumn that is up to 1.2 inch (3 cm) long. Numerous anthers branch out from the midcolumn in the upper two-thirds. The pistil is a little longer than the midcolumn; the stigmas are white or yellow.
Fruits: Fruits are flattened and spherical and up to .8 inch (2 cm) long and 1.8 inch (4.5 cm) thick. When ripe, they are black and do not open on their own. Seeds are up to .4 inch (1 cm) long.
Occurrence: Frequent in coastal regions, often on the land side of mangrove forests, cork trees are also planted as street trees.
Other names: English "Milo," "Pacific Rosewood," Portia Tree"; German "Pappelblättriger Eibisch"; Spanish "Cremón."
Additional information: The seeds of the cork tree can survive for one year in sea water and then bud on hot beaches. They have reached almost all tropical coasts. The Polynesians first brought the tree to Hawaii. The wood is cream white to chestnut brown to almost black. Because it is easy to polish, the grained wood is used to make receptacles of all kinds. The flowers of the very similar coast hibiscus (see page 170) have a darker center, a strongly split epicalyx, and purple stigmas.

Potato Tree

Solanum wrightii
Family: *Solanaceae*, Solanum

Most important features: The leaves are lobed. The midrib at the bottom side is often lined with prickles. The flowers begin as striking violet, but the color fades rapidly. At the end, the flowers are almost white with a yellow stamen cone.
Growth structure: The potato tree may grow up to 40 feet (12 m) tall, but it is often considerably smaller. It frequently has thorns on the trunk and branches.
Leaves: The leaves are alternating. They are 3.6 to 15.8 inches (9 to 37 cm) long and 2 to 12.8 inches (5 to 32 cm) wide. The leaves are ovate to elliptical with five to nine more-or-less distinct lobes. They are lighter on the bottom side; the upper side has more hair.
Flowers: The flowers grow in erect panicles that are 2 to 3 inches (5 to 7.5 cm) wide with five green, hairy sepals and five petals. At the edge, the flowers are wavy.
Fruits: The spherical fruits are berries. They are greenish yellow to orange and 1.2 to 2 inches (3 to 5 cm) in diameter.
Occurrence: One finds potato trees used as ornamental trees in parks in South America. Originally, they were from Brazil.
Other names: German "Kartoffelbaum"; Spanish "Azulillo," "Arbol de Papa."
Additional information: When they are older, the almost white flowers of the potato tree look like overly large potato flowers. Indeed, the tree is related to the potato (see also pages 150 and 198); however, it doesn't develop tubercles. It grows rapidly and often reaches its final height after only two to four years. Most other species of *Solanum* remain herbaceous, and many smaller species have very similar leaves. This is also true of the quito tomato (*S. Quitoense*) which is quite striking because of its violet hairs. Its orange berries are edible, like the fruits of the related eggplant (*S. Melongena*). The other fruits of this genus are extremely poisonous.

Gustavia

Gustavia augusta
Family: *Lecythidaceae*, Lid pot trees

Most important features: The leaves are large and accumulate at the ends of the branches. The large flowers have six to nine white petals and many stamens connected at the base. The stamens are white at the base and pink to pale lilac toward the end. The anthers are yellow.
Growth structure: Gustavia grows up to 65 feet (20 m) high. It is usually smaller with firm branches that often start to grow from the ground up.
Leaves: The leaves are alternating and up to 20 inches (50 cm) long, four times longer than their width. They are often slightly wavy with many lateral and sublateral veins. The leaves may have an indistinct dentate margin.
Flowers: These often grow in simple racemes with few blooms. They are 4 to 8 inches (10 to 20 cm) long with a firm petiole that turns into a bowl-shaped calyx consisting of four indistinct lobes. The flowers are white to light pink and have a pleasant odor.
Fruits: The fruits are brown, almost spherical, and up to 3.2 inches (8 cm) in diameter. At the top, they are flat to concave.
Occurrence: Gustavia grows in the Western Hemisphere and occasionally in Asia. Originally it was from the northern part of South America.
Other names: German "Gustavia"; Spanish "Guatoso," "Matamatá"; Portuguese "Geniparaná."
Additional information: Gustavia is not used as an ornamental very often. However, in full bloom it is quite impressive. Several of the forty-one species of the genus are cultivated; some are very difficult to differentiate. Membrillo (*G. superba*) has spread from Costa Rica to the northern part of Colombia. However, one can even find it in eastern Asia. It has fewer branches and larger leaves, often up to 4 feet (1.3 m) long. Although also called stinkwood, it is not related to the South African stinkwood (*Ocotea bullata*), now almost extinct because of the heavy demand for its wood for furniture.

Elephant Apple

Dillenia indica
Family: *Dilleniaceae*, Roseapple plants

Most important features: The trunk is red brown. The leaves have many firm, lateral veins. These run almost parallel to the dentate margin of the leaf. The white flowers are very large.
Growth structure: The elephant apple grows up to 50 feet (15 m) tall with a short trunk and a wide, dense top.
Leaves: The leaves are alternating and tapering. They grow up to 15 inches (35 cm) long. The upper side shimmers. The leaves have up to forty pairs of veins.
Flowers: The flowers are up to 8 inches (20 cm) long with five thick sepals that are spherical and green. The five large, asymmetric petals are slightly wavy. The flowers have a multitude of yellow stamens and twelve to seventeen ray flowers. The flowers only live for one day.
Fruits: These are shaped like apples. They grow up to 6 inches (15 cm) in diameter. The fruits are green and have overlapping, fleshy sepals. The individual, slimy carpels are hidden in the sepals.
Occurrence: Originally from Southeast Asia, the elephant apple is rarely found in other regions.
Other names: English "Indian Roseapple"; German "Indischer Rosenapfel."
Additional information: Because the Indian rose apple (not to be confused with the rose apple, see page 92), has a very pleasant odor that is similar to apples, many animals like to eat it. One can make a refreshing drink from the fleshy sepals by mixing them with water and sugar. Some people cook them as vegetables or make them into curries. The Philippine dillenia (*D. philippinensis*, see photo at left), which also has large flowers with red violet stamens and ray flowers, is used in the same way. Its sepals are smaller at the fruit, but the fleshy petals are larger.

Durian

Durio zibethinus
Family: *Bombacaceae*, Cotton tree plants

Most important features: The large flowers are cream colored and hang in dense bunches from the trunk and stronger branches. The gigantic fruits are covered with wide, angular thorns.
Growth structure: This tree grows up to 130 feet (40 m) high. The branches are almost horizontal.
Leaves: The leaves are alternating and up to 12 inches (30 cm) long and 6 inches (15 cm) wide. They are tapering, elliptical, and pointed. On the upper side, they are bright and shimmering; on the bottom side, they have subtle tiny scales.
Flowers: The flowers grow up to 4 inches (10 cm) long. The calyx is bell-shaped with four to six dentate petals and numerous stamens extending out from the flower. The pistil is even longer.
Fruits: The fruits are gray green and spherical. They grow up to 12 inches (30 cm) long and weigh up to 17 pounds (8 kg). The fruits have five parts. The inside is almost white. Each section has two to six large seeds.
Occurrence: Durian is grown in many gardens in Southeast Asia, but rarely in other areas.
Other names: English "Doorian"; German "Durian," "Stinkfrucht"; Portuguese "Durião"; Spanish "Durión."
Additional information: No other fruit elicits such a range of emotions. The smell is so bad and so penetrating that one can only carry it on public transportation in airtight containers. The smell is enough to make many people drowsy. On the other hand, the coat of seeds is considered one of the greatest delicacies in Southeast Asia. Often it is eaten uncooked; however, there are other ways to prepare it, including durian ice cream. Because the fruits spoil so rapidly, they are rarely seen outside of Southeast Asia.

Lime

Citrus aurantiifolia
Family: *Rutaceae*, Rue plants

Most Important Features: The leaves have pointed thorns with a clear segment between the leaflets and the slightly widened petioles. The edges have little scallops. The flowers are white, and the spherical fruits are green.
Growth structure: This is a small tree or high bush that grows up to 16 feet (5 m) high. The top is dense and rounded.
Leaves: The leaves are alternating, elliptical, shimmering, and up to about 4 inches (10 cm) long.
Flowers: These grow individually or in groups. They are about 1.2 to 1.6 inch (3 to 4 cm) in diameter with a wonderful odor. The flower has a green calyx, four to six white petals, numerous stamens, and a stigma shaped like a head.
Fruits: The fruits are spherical and 1.2 to 2.8 inches (3 to 7 cm) in diameter. When ripe, they are green or greenish yellow. Inside, the fruits look like lemons.
Occurrence: One often finds limes growing in yards. Originally, they were from the Indo-Malaysian area.
Other names: German "Limette," "Limone"; Spanish "Lima," "Limón Criollo"; Portuguese "Lima."
Additional information: The lime is the most popular citrus fruit in tropical gardens. The juice and the oil of the peel are used in numerous dishes, drinks, and medicines. The juice has much more flavor than lemon juice. The shaddock or pummelo (*C. maxima*) is often cultivated in the humid tropics. Its petioles are as wide, and its pear-shaped fruits are much larger.

Blue Gum

Eucalyptus globulus
Family: *Myrtaceae*, Myrtle plants

Most important features: When bruised or injured, the leaves emit an odor like cough drops. Young leaves are bluish and waxy. Flowers have a multitude of cream white stamens. The perianths appear to be falling out of the flowers.
Growth structure: Blue gum trees grow up to about 65 feet (20 m) tall. In their native habitat, however, they grow up to 230 feet (70 m) high. The gray trunk is erect, and the bark comes off in long brown strips.
Leaves: Leaves are very variable. In young plants (photo at left), they are opposite, heart-shaped, and up to 6 inches (15 cm) long and 4 inches (10 cm) wide. Older leaves (photo at right) are alternating and up to 15 inches (35 cm) long and 1.2 inch (3 cm) wide. The petioles are oblong.
Flowers: These are up to 1 inch (2.5 cm) long with a conical base. They grow individually or in clusters of up to seven.
Fruits: The capsules, with semispherical bases, are up to 1 inch (2.5 cm) long and often have four longitudinal ribs. They are flatly vaulted with three to five openings.
Occurrence: Found in the tropics and subtropics, they originally came from the southern part of Australia and Tasmania.
Other name: English "Tasmanian Blue Gum"; German "Blauer Eucalyptus," "Fieberbaum."
Additional information: The blue gum is representative of more than six hundred *Eucalyptus* species. Almost all come from Australia, and even professionals struggle to tell them apart. Most species have white blooms, though some have yellow or red. Because many species grow very rapidly, they are planted all over the world to supply lumber for the paper industry, often driving out native vegetation. Robust species can thrive in areas with mild winters. *E. camaldulensis*, with white bark, is also frequently planted. Its red core wood resists termites, as does the only species found in the Philippines, *E. deglupta* (Mindanao gum), whose bark has lengthy stripes.

Cashew Nut

Anacardium occidentale
Family: *Anacardiaceae*, Sumach plants

Most important features: The leaves are obovate with light, relatively obtuse lateral veins. The flowers are small, whitish, and pink. The fruits have two parts (see below).
Growth structure: The cashew nut tree is usually small, seldom growing higher than 50 feet (15 m). Often, it is only a bush.
Leaves: The leaves are alternating. They grow up to 8 inches (20 cm) long and only half as wide. The leaves are leathery and bare with a continuous margin.
Flowers: These grow in richly branched inflorescences at the ends of the branches. The flowers have five petals that change colors in the flowering season from greenish white to pink.
Fruits: The fruits are divided into two parts and hang down. One part is up to 4 inches (10 cm) long. It is fleshy and shaped somewhere between an apple and a pear. When ripe, it is yellow to red and grows from the petiole. The much smaller part is only about 1 inch (2.5 cm) long, kidney-shaped, and bent. It is brown with a hard peel that covers the seeds.
Occurrence: Cashew nut trees grow in all tropical areas. They originated in Central America, the Caribbean, and the northern part of South America;
Other names: English "Cashew Apple"; German "Cashewnuss," "Kaschubaum"; French "Acajou"; Portuguese "Cajú"; Spanish "Merey," "Marañon."
Additional information: The cashew nut tree has a variety of uses. In earlier times, people ate the fleshy fruit stalk. The nut was more or less ignored because its shell contains a poisonous oil that irritates the skin. However, this was used in folk medicine. Today, the oil is used to produce paints, varnish, and brake and clutch linings. After roasting, the pit becomes the delicious cashew nut that is forty-five percent fat and twenty percent protein.

Mango

Mangifera indica
Family: *Anacardiaceae*, Anacardia plants

Most important features: The leaves are slender and dark green with up to thirty pairs of yellow lateral veins that branch obtusely. The small flowers are numerous and hang in richly branched panicles. The fruits are large and hang on long petioles.
Growth structure: These trees grow up to 100 feet (30 m) high with dense, rounded tops.
Leaves: The shimmering leaves are alternating and up to 12 inches (30 cm) long. The young leaves are pale or reddish and droopy.
Flowers: The flowers are up to .2 inch (5 mm) long. Usually, they are yellowish, but they may be cream white to pink. Often thousands grow on branches.
Fruits: The fruits vary widely, depending on the specific type, but they are always slightly crooked. Often, they are approximately 4 inches (10 cm) long, but some may be up to 16 inches (40 cm) long. The shape ranges from almost spherical to tapering to kidney-shaped. The fruits start out a bluish green, but later, they are yellow, orange, reddish, or violet. The fibrous flesh is yellow to orange and contains a large pit.
Occurrence: Mangos grow naturally in many tropical areas. They are cultivated in the Mediterranean region. Originally, they were from India.
Other names: Mango is the name in almost all languages.
Additional information: The mango is among the most important of all tropical fruits. It was mentioned in writings more than four thousand years ago. It is a highly nutritious food, containing sugar, vitamins, and proteins. Some people are allergic to the resin of the mango tree. In Southeast Asia, other species are used, especially the bachang (*M. foetida*), whose fruits have a pleasant, slightly sour taste but an unpleasant odor.

Sea Grape

Coccoloba uvifera
Family: *Polygonaceae*, Knotgrass plants

Most important features: The trunk is crooked. The leaves are wide and spherical with a short tube at the base of the petiole. The flowers and fruits grow in long, slender clusters.
Growth structure: The sea grape grows up to 50 feet (15 m) tall with a very short trunk and a wide top of branches that are wound together.
Leaves: The leaves are alternating and very widely spherical with a short petiole. They grow up to 10 inches (25 cm) in diameter with a heart-shaped base. The leaves are leathery with many veins. Old leaves are often reddish.
Flowers: The flowers are small and cream white. At first, they grow in erect simple racemes that are up to 12 inches (30 cm) long.
Fruits: These are spherical and up to .8 inch (2 cm) in diameter. The fruits are red to dark purple, waxy, and grow in long, hanging simple racemes that are up to 16 inches (40 cm) long. The glassy, sweet-sour flesh is produced by the perianth, the tips of which are still recognizable. Inside is a light, hard, ovate nut.
Occurrence: One often finds sea grape trees growing on beaches, but one can also find them in gardens. Originally, they were from the coasts of the Western Hemisphere.
Other names: German "Seetraube," "Meertraube," "Strandtraube"; French "Raisin Bord de Mer"; Spanish "Uvero de Playa."
Additional information: The sea grape can withstand salt and strong wind. On sandy beaches, one finds the tree just far enough away from the waterline that the rain can wash the salt off of its roots. Juice, jelly, and even wine are made out of the fruits. With about 120 species, the *Coccoloba* genus can be found almost everywhere in the tropics of North and South America. The grand leaf sea grape (*C. pubescens*) is also cultivated; its enormous leaves are often more than 3 feet (1 m) in diameter.

Cocoa

Theobroma cacao
Family: *Sterculiaceae*, Cacao plants

Most important features: The flowers grow on the trunk and the thicker branches. They are small and grow in groups of five. They are cream white with purple. The fruits are large with longitudinal bulges.
Growth structure: Cocoa trees grow up to 50 feet (15 m) tall. They have a short trunk that is kept shorter when cultivated.
Leaves: Alternating leaves are up to 16 inches (40 cm) long, oval to tapering, pointed, and dark green. Young leaves are pale green or pink.
Flowers: The flowers grow up to .6 inch (15 mm) in diameter. The sepals up to .3 inch (8 mm) long. The petals are arranged into a hood-shaped lower part that is streaked with pink. They have a petiole and a cream-colored upper part. The five long, purple filaments are sterile.
Fruits: Fruits grow up to 12 inches (30 cm) long, varying from spherical to ovate. The base may be obtuse or acute. Fruits are yellow to brown red when ripe. The peel is about .8 inch (2 cm) thick. Inside, the flesh is white and contains numerous seeds up to 1.2 inch (3 cm) long.
Occurrence: Often cultivated on plantations, the cocoa came originally from the Amazon region.
Other names: German "Kakao"; French and Spanish "Cacao"; Portuguese "Cacau."
Additional information: Cocoa was cultivated in Central and South America long before the arrival of Europeans. For the Aztecs, it was the food of the feathered snake god, Quetzalcoatl. It was used as a means of payment for war and as a stimulant during war. Casanova considered it a love potion. Actually, cocoa contains stimulating ingredients that can cause a kind of addiction. Right after the harvest, the seeds (cocoa beans) are undrinkable; they have to be fermented to reduce the bitterness and to build up the typical aroma. The fat (cocoa butter), which is a by-product of cocoa powder, is used in cosmetic and pharmaceutical products.

Chicle

Manilkara zapota
Family: *Sapotaceae*, Sapote plants

Most important features: All parts of the tree have a milky sap. The leaves accumulate at the ends of the branches. The small, white flowers have a brown, hairy calyx that is often hidden. The fruits are a dull brown and spherical.
Growth structure: Chicle trees grow up to 100 feet (30 m) high, although they are often much smaller. The trees have a short trunk and a dense top.
Leaves: The dark green leaves are alternating to almost opposite with indistinct lateral veins. They are tapering and up to 6 inches (15 cm) long.
Flowers: Each flower grows individually in a bract. The flowers are up to .6 inch (1.5 cm) long. They grow in groups of six. The buds are covered by three outer sepals. The flowers have a thick pistil.
Fruits: The fruits are spherical to oval and slightly rough. They are usually about 2 inches (5 cm) long, but they may grow up to 4 inches (10 cm) long. The flesh is a glassy yellow brown with up to twelve (more often, three) black, flattened seeds.
Occurrence: Chicle is cultivated in all tropical areas. Originally, it was from Mexico and Central America.
Other names: English "Naseberry," "Sapodilla"; German "Breiapfel," "Sapote"; Portuguese "Sapoti"; Spanish "Chicozapote," "Nispero."
Additional information: The flesh of the chicle is sweet, aromatic, and mushy. It has little grains like a soft pear. Often one uses a spoon to eat it out of the inedible peel. In earlier times, chicle was cultivated for its milky sap, drawn out of its scratched bark. The juice supplied the basic material for chewing gum. Today, this is produced synthetically. The closely related bullet wood (*Manilkara bidentata*) has a stronger latex that is used as the cover of golf balls.

Barbados Almond

Terminalia catappa
Family: *Combretaceae*, Almond plants

Most important features: The top grows out of clearly separated levels of horizontal branches. The fruits are almond-shaped.
Growth structure: The Barbados almond grows up to 80 feet (25 m) high, but often is smaller.
Leaves: The leaves grow in dense bunches on short parts of the branches that are bent upward. The leaves are up to 15 inches (35 cm) long and 10 inches (25 cm) wide and obovate to spherical. They are pointed, stiff, and leathery with firm light green veins. Before the leaves drop, they turn red.
Flowers: The flowers are small and white with five star-shaped corollas and ten filaments in spikes up to 10 inches (25 cm) long.
Fruits: These are oval, flattened, stony fruits that grow on a clear, narrow limb. They are up to 2.8 inches (7 cm) long, 2 inches (5 cm) wide, and 1 inch (2.5 cm) thick. For a long time, they are bluish green. However, when they are ripe, they sometimes turn yellowish to wine red.
Occurrence: Frequently planted along streets, especially close to the coast, the Barbados almond is originally from the coasts of the Indian Ocean to the western Pacific Ocean.
Other names: English "False Kamani," "Indian Almond," "Tropical Almond"; German "Strandmandel," "Indischer Mandelbaum," "Katappenbaum"; French "Amandier des Indes," "Amandier Tropical"; Spanish "Alconorque," "Almendra de India," "Almendrón."
Additional information: The fruits of the Barbados almond float and are spread by the ocean. The sour flesh and almond-shaped content of the pit are edible. The fruits of the closely related billygoat plum (*T. ferdinandiana*) contain about fifty times more vitamin C than oranges. The fruits of species such as *T. chebula* are less tasty and are usually used for tanning. The wood of *Terminalia superba* is very popular for interior design.

Soursop

Annona muricata
Family: *Annonaceae*, Scale apple plants

Most important features: The leaves are dark green and grow in two vertical rows. On the upper side, they are shimmering; on the bottom side, they are dull. The flowers are thick and fleshy with heart-shaped petals. The fruits are very large and green with many soft prickles.
Growth structure: The soursop is a small tree, rarely reaching 33 feet (10 m) in height.
Leaves: The evergreen leaves are alternating and tapering. They grow up to 8 inches (20 cm) long with a continuous margin.
Flowers: These grow individually or in a small group. They are greenish yellow. The petals are up to 2 inches (5 cm) long and almost as wide. Three petals grow together in two circles. The stamens and carpels are small, numerous, and densely packed.
Fruits: The fruits are spherical to tapering and up to 15 inches (35 cm) long. They weigh up to 13 pounds (6 kg). However, they are often somewhat smaller. The fruits are covered with prickles that are about .4 inch (1 cm) long. The flesh is white and somewhat sour. The seeds are black and up to .8 inch (2 cm) long.
Occurrence: Widely cultivated, the soursop originated in the Western Hemisphere.
Other names: German "Stachelannone," "Sauersack"; Spanish "Guanábana."
Additional information: The soursop is among the tastiest of tropical fruits. Only a few of the 137 species are cultivated. Unfortunately, because the fruits are very sensitive to pressure, exporting them is a problem. Of all the species, the soursop has the largest fruits. Because of the weight of the fruits, the flowers grow out of the trunk or strong branches. The seeds of all the members of this genus are poisonous; however, they are so large that no one would swallow them by mistake.

Sweetsop

Annona squamosa
Family: *Annonaceae*, Scale apple plants

Most important features: The leaves grow in two vertical rows. They are light green, but they are even paler at the bottom of the sides. The flowers have three oblong petals. The fruits are as large as an orange and are marked with bluish green spots.
Growth structure: The sweetsop is a bush or a small tree. It often grows 16 to 23 feet (5 to 7 m) high; occasionally, it will grow up to 40 feet (12 m) tall.
Leaves: The leaves, alternating and tapering., are up to 6.8 inches (17 cm) long with a continuous margin. The leaves are a dull green and slightly hairy. The leaves drop in the dry period.
Flowers: These grow individually or in small groups. They are greenish yellow with slender petals up to 1.2 inch (3 cm) long. The stamens and carpels are small, numerous, and densely packed.
Fruits: The fruits are spherical and made up of many segments. They are usually green, blue, or violet. They are waxy with a whitish, sweet flesh. The seeds are black and about .4 inch (1 cm) long.
Occurrence: The sweetsop tree is widely cultivated. It was originally from the Western Hemisphere.
Other names: English "Sugar Apple"; German "Schuppenannone," "Süßsack," "Rahmapfel"; Portuguese "Fruta do Conde"; Spanish "Riñon."
Additional information: Of all the members of this genus, this species is probably seen most often. Among the varieties are those with violet fruits or without seeds. Among the finest is the cherimoya (*Annnona cherimola*). Its fruit remains green and is not deeply divided. It only has flat scales and is very sweet. The custard apple (*Annona reticulata*) comes from the hot lowlands. The plant looks like the soursop, and its fruits turn red when they are ripe. They are only a little wavy or unclearly divided into mounds.

Bread Fruit

Artocarpus altilis
Family: *Moraceae*, Mulberry plants

Most important features: This tree has small, firm branches. The leaves are very large and deeply divided. All parts of the tree have a milky sap. The large fruits are thorny.
Growth structure: Often a small tree that only reaches 33 feet (10 m), the bread fruit tree can grow up to 100 feet (30 m) tall. The top is wide and irregular.
Leaves: The leaves are alternating and shimmering green. They grow up to 35 inches (90 cm) long and 20 inches (50 cm) wide with yellow veins on both sides. Each leaf is separated into five to ten acute lobes.
Flowers: The tiny flowers grow in large inflorescences. The male inflorescences are piston-shaped and hanging. They grow up to 8 inches (20 cm) long. The female ones are spherical and rather erect when they are young.
Fruits: The fruits are spherical to tapering and up to 12 inches (30 cm) long. They are green, but turn yellow when they are ripe. They have soft thorns with large seeds up to 1.2 inch (3 cm) long, or they have wart-shaped thorns and no seeds.
Occurrence: The bread fruit trees grow in all tropical areas. Scientists assume they originated in Malaysia.
Other names: German "Brotfruchtbaum"; Spanish "Arbol de Pan."
Additional information: About twenty percent of the fruit of the bread tree is starch. Another one to two percent is protein. The unripe fruit can be cooked, fried, stewed, baked, or dried for long-term storage. The seeds of the ripe fruits (bread nuts) are eaten roasted. In the Pacific region, the bread fruit is a dietary staple. In South America, it was as much of a staple as the banana (see page 50). In 1789, Captain Bligh of the *Bounty* picked up one thousand young bread fruit trees from Tahiti to bring to South America to feed the slaves. This first attempt resulted in the famous mutiny; however, in 1793, Bligh's second attempt succeeded.

Chempedak

Artocarpus integer
Family: *Moraceae*, Mulberry plants

Most important features: Leaves are dark green with yellow veins. All parts of the tree have a milky sap. Large fruits have delicate warts or dull thorns and grow on trunk and branches.
Growth structure: This tree grows up to 65 feet (20 m) high with a firm main trunk and an elliptical top.
Leaves: The leaves are alternating, simple, and elliptical. They are 2 to 10 inches (5 to 25 cm) long, 1 to 4.8 inches (2.5 to 12 cm) wide, leathery, and pointed. At the bottom, they are clearly separated from the petiole.
Flowers: The tiny flowers grow in large inflorescences. The male inflorescences are cylindrical to club-shaped and 1.2 to 2.2 inches (3 to 5.5 cm) long; the female ones are often more spherical and up to 2.4 inches (6 cm) long.
Fruits: The fruits are elliptical and 8 to 15 inches (20 to 35 cm) long and 4 to 6 inches (10 to 15 cm) thick. They are green, but they turn yellowish when they ripen. The seeds are up to 1.2 inch (3 cm) long.
Occurrence: Now pan-Tropical, Chempedak was originally from the region around Malaysia.
Other names: The Malayan name "Chempedak" is used by all other languages.
Additional information: In Southeast Asia, one finds more chempedak than bread fruit trees (see above). In other tropical areas, one often finds the very similar Jack fruit (*A. heterophyllus*). By the sixteenth century, the Portuguese had brought it to Brazil where the fruits grow in clusters up to 3 feet (1 m) long, weighing up to 110 pounds (50 kg). The fruits are more fibrous and the odor more unpleasant. However, the odor can be removed by placing them into salt water. In Thailand, the Jack fruit tree is thought to bring luck. The yellow dye used for the hoods of Buddhist monks is made from its wood. Southeast Asians enjoy the wonderful smelling marang (*A. odoratissimus*), which has larger, dull leaves and yellow, aromatic fruits.

Banyan

Ficus benghalensis
Family: *Moraceae*, Mulberry plants

Most important features: This is a widely sweeping tree with numerous aerial roots that often grow into new trunks. All parts of the tree have a sticky, milky sap. The leaves are hairy on the bottom side.
Growth structure: The banyan tree grows up to 80 feet (25 m) high, often with several trunks. The trunk continues to widen.
Leaves: The leaves are alternating and oval. They grow up to 10 inches (25 cm) long. They are stiff, leathery, and shiny on the upper side.
Flowers: As with all figs, the flowers are tiny and hidden in a fruitlike formation.
Fruits: The fruits are spherical and up to .8 inch (2 cm) long. They are yellow for a long time, but turn red when they are ripe.
Occurrence: Banyan trees are found in tropical Asia. They originated in India.
Other names: English "East India Fig"; Spanish "Laurel de India."
Additional information: The name "banyan" is derived from the Indian word for merchants, who spread their goods in its shadow. Because the branches are supported by trunks that are always growing thicker, the banyan tree can cover immense areas. The largest known tree covers about 5 acres (2 hectare). Like many of the approximately 750 *Ficus* species, the banyan is a strangler fig as well. Its seeds thrive on other trees, from which the young plants send aerial roots toward the ground. When those reach the ground, they rapidly turn into trunks. The dense coat of banyan trunks eventually crushes the host tree. The Hindus consider the banyan tree to be sacred because it is mentioned in the history of several gods. Hindus and Buddhists think even more highly of the peepul tree or sacred fig (*F. religiosa*) under which Buddha is supposed to have reached his illumination. This tree is easily recognizable because of the long petioles and leaves that are often bluish green and heart-shaped as well as acute.

Indian Rubber Fig

Ficus elastica
Family: *Moraceae*, Mulberry plants

Most important features: The Indian rubber fig produces a sticky, milky sap. The leaves are large, elliptical, and smooth. The youngest leaves are curled to a long pointed end and covered by a reddish, skinlike hull.
Growth structure: This tree grows up to 100 feet (30 m) high, but it is frequently smaller. The trunk is short and usually covered in numerous, thickened aerial roots. The branches are often split for several feet (meters).
Leaves: The leaves are alternating and up to 15 inches (35 cm) long. They are stiff, leathery, and pointed with strong main veins and very delicate lateral and sublateral veins.
Flowers: As with all figs, the flowers are tiny and hidden in a fruitlike formation.
Fruits: The fruits are up to .8 inch (2 cm) long and ovate with a cone-shaped thickened petiole. They are faintly reminiscent of an acorn. When ripe, they are reddish. They often grow in pairs in the bracts and are rarely seen.
Occurrence: Often planted as a shade tree, the Indian rubber fig originated from Assam to Java.
Other names: English "Assam Rubber Tree"; German "Gummibaum"; Spanish "Caucho," "Palo de Goma."
Additional information: Up to the beginning of this century, the Indian rubber tree was grown on plantations to make latex from its milky sap. Today, the rubber tree (*Hevea brasiliensis*, see page 86) has taken its place. However, it is still one of the most popular indoor plants. The same is true of the Benjamin tree (*F. benjamina*), which is distinguished by its drooping branches and much smaller leaves. Another species, the fiddle leaf fig (*F. lyrata*), can be often found in our living rooms. Its leaves grow up to 24 inches (60 cm) long and have firm lateral veins, a heart-shaped to ovate base, and a widely spherical tip.

Sensitive Plant

Mimosa pudica
Family: *Mimosaceae*, Mimosa plants

Most important features: When touched, the pinnules curl, and the leaves fold down. The flowers are pink to light violet and grow in small, spherical, sessile clusters.
Growth structure: This is a subshrub that is slightly woody and about 3 feet (1 m) high with thorny branches.
Leaves: The leaves are alternating and bipinnately divided with one or two paired pinnas of the first order. These start almost like fingers from one point. Each has ten to twenty-six tapering, pinnules that are .3 inch (8 mm) long and only half as wide.
Flowers: Many flowers grow together in .4 to .8 inch (1 to 2 cm) wide clusters. The stamens of the tiny individual flowers are .2 to .3 inch (5 to 8 mm) long.
Fruits: Several fruits grow together in one bunch up to 1 inch (2.5 cm) long and .2 inch (5 mm) wide. The fruits are flat with a bristly, hairy frame. When the fruit is ripe, the seed falls out.
Occurrence: These plants grow as weeds in all tropical areas. Originally, they were from the Western Hemisphere.
Other names: English "Shame Plant," "Touch-me-not"; German "Mimose," "Sinnpflanze"; Spanish "Dormidera," "Sensitiva."
Additional information: Not only does the sensitive plant move, but almost all plants do as well. However, this happens so slowly that it is only clear when seen with a time-lapse camera. Only with the sensitive plant and a few other plants can one clearly observe movements with the naked eyes. In this case, touching the plant sends a signal, which is transported .2 to 4 inches (0.5 to 10 cm) per second, depending on the temperature and strength of the stimulus. Certain cells in the leaf joints react to the signal by changing their inner pressure, shrinking, and causing movement.

Cockspur Coral Bean

Erythrina crista-galli
Family: *Fabaceae*, Butterfly flowers

Most important features: The leaves of this thorny plant grow in threes. The flowers are bright red with one large, wide petal at the bottom and a bent, tubular structure above it.
Growth structure: The cockspur coral bean usually grows as a wide bush with a thick, wooden base. However, as a tree, it can grow up to 30 feet (9 m) tall with a thick trunk.
Leaves: Leaves are alternating with thorns at the petiole and the midrib. The small leaves are tapering, elliptical, and up to 6 inches (15 cm) long. The leaves may drop during the dry period.
Flowers: The flowers often grow in simple racemes at the ends of the branches. The two upper petals form a tube that is almost closed around the stamens and pistil. The two petals on the side are very small. The one at the bottom is up to 2.4 inches (6 cm) long and 1.4 inch (3.5 cm) wide. Often, this petal is slightly boat-shaped and bent toward the ground.
Fruits: The fruits are bean-shaped and grow up to 12 inches (30 cm) long and .8 inch (2 cm) thick.
Occurrence: The cockspur coral bean grows in the tropics and subtropics. It originated in South America.
Other names: English "Cry Baby"; German "Korallenstrauch."
Additional information: Like most of the *Erythrina* species (see page 76), the cockspur coral bean is pollinated by birds. In order to attract them, the plant produces so much nectar that it often drips out of the flowers. For this reason, the plant is also referred to as "cry baby." The bush can withstand temperatures as low as 23°F (–5°C). Although the younger branches often freeze, the plant sprouts again and again from its rootstock. The South African dwarf coral tree (*E. humeana*) is often grown as a bush in a pot. It differs from the former because of its triangular to three-lobed leaves and because the smaller, much more slender flowers grow in erect inflorescences.

Red Powder Puff

Calliandra haematocephala
Family: *Mimosaceae*, Mimosa plants

Most important features: Leaves branch into two axils, each with three to ten pairs of leaflets. Inflorescences, shaped like powder puffs, are 2 to 3.6 inches (5 to 9 cm) long and often crimson.
Growth structure: This low, narrow bush, or small tree, rarely reaches 16 feet (5 m) high.
Leaves: Alternating leaves grow up to 8 inches (20 cm) long. Leaflets grow larger toward the tip and may be .8 to 5.6 inches (2 to 14 cm) long and .2 to 2.4 inches (0.5 to 6 cm) wide, varying in shape from crooked elliptical to tapering ovate.
Flowers: The flowers are very small and grow twenty-five to eighty-five in dense clusters. While still a bud, the flower is reminiscent of an unripe blackberry. The many filaments, which are up to 1.8 inches (4.5 cm) long, create the powder puff. All other parts of the flower are rather insignificant in comparison.
Fruits: The fruits are bean-shaped, 2.4 to 5.2 inches (6 to 13 cm) long, and about .4 inch (1 cm) wide. They are flat with a thickened margin.
Occurrence: Frequently used as an ornamental bush, the red powder puff grows wild in many regions. Originally, it was from Bolivia.
Other names: German "Roter Puderquastenstrauch"; French "Pompom"; Spanish "Bellota," "Flor de la Cruz."
Additional information: The name "red powder puff" is used for all *Calliandra* species that have red, semispherical inflorescences. Sometimes, one can find plain white or pink white inflorescences among the same species. *C. tergemina var. emarginata* (see photo below at right) has one pair of pinnately divided leaves at the branched leaf and one significantly larger leaf at each of the two forked branches. In Southeast Asia, the red species is more common; in the Western Hemisphere, the red white species is. In subtropical areas with a risk of frost, the *C. tweediei* (see photo below at left) is preferred. It has much more delicate, bipinnately divided leaves and only eight to sixteen flowers per cluster.

Pink Powder Puff

Calliandra surinamensis
Family: *Mimosaceae*, Mimosa plants

Most important features: The leaves branch in two axils, each with seven to twelve pairs of leaflets. The inflorescences look like powder puffs. Inside, they are white with a few white flower tubes. They are pink outside. The fruits grow vertically toward the top.
Growth structure: This is a low, wide bush that rarely grows more than 7 feet (2 m) high. Typically, the branches are almost horizontal.
Leaves: The leaves are alternating and up to 4 inches (10 cm) long. The leaflets are narrow and .4 to .8 inch (1 to 2 cm) long. They are shiny with a clear bulge at the base.
Flowers: The flowers are tiny and grow in dense clusters. The powder puff effect is created by numerous filaments that grow up to 4 cm long.
Fruits: The fruits are bean-shaped, 2 to 3.2 inches (5 to 8 cm) long, and up to .4 inch (1 cm) wide. They are rather flat with a thickened margin.
Occurrence: One finds the pink powder puff in gardens. It originated in the northern part of South America.
Other names: German "Rosa Puderquastenstrauch"; French "Pompom de Marin."
Additional information: The inflorescences of the pink powder puff are reminiscent of the ones of the rain tree (see page 64), yet the growth structure is entirely different. About eighty of the approximately 130 species of the *Calliandra* genus have spherical flower clusters (see also photos at top and bottom right); the remaining fifty, on the other hand, have oblong inflorescences. These can be very decorative. For example, the angel hair bush (*C. calothyrsus*) has filaments that are up to 2.8 inches (7 cm) long. The filaments are often crimson with a greenish yellow perianth. Unfortunately, they wilt very quickly, so the full splendor is only noticeable in the early morning hours. The angel hair bush is occasionally cultivated to improve the soil and as a forage crop for bees and shellac-producing insects.

Barbados Flower

Caesalpinia pulcherrima
Family: *Caesalpiniaceae*, Carob plants

Most important features: The leaves are bipinnately divided. The flowers grow in erect simple racemes that are often bright red. The flowers have five petals. The upper one is coiled and often a different color than the other four. The pistil and ten stamens stand out up to 4 inches (10 cm) beyond the flower.
Growth structure: As a bush, the Barbados flower grows up to 10 feet (3 m) high. As a small tree, it may reach as high as 20 feet (6 m). The tree has a loose top with thorns on the trunk.
Leaves: The leaves are alternating and up to 12 inches (30 cm) long. They have three to nine feathers of the first order. Each of these has six to twelve pairs of small leaflets up to .4 to 1.2 inch (1 to 3 cm) long.
Flowers: The flowers grow on long petioles. The stamens and pistil are bent slightly upward. The simple racemes are up to 16 inches (40 cm) long.
Fruits: The fruits are bean-shaped, up to 4.8 inches (12 cm) long, and .8 inch (2 cm) wide. They are flattened. When ripe, they are black brown and burst into two parts.
Occurrence: Barbados flower is planted in gardens in all tropical areas; it also grows wild. Its place of origin is uncertain, but the Antilles is a possibility.
Other names: English "Dwarf Poinciana," "Peacock Flower," "Pride of Barbados"; German "Stolz von Barbados," "Pfauenstrauch," "Zwergpoinciane"; French "Petit Flamboyan," "Orgueil de Chine"; Spanish "Clavellina."
Additional information: The Barbados flower often has bright red petals. Some of the flowers have a yellow margin. However, there is also a plain yellow form. The leaves are poisonous, but they are supposed to have a laxative and antipyretic effect. Tanning substances come from many of the closely related species, and a red coloring agent comes from the sappan wood (*C. sappan*).

Trumpet Flower

Tecoma stans
Family: *Bignoniaceae*, Trumpet tree plants

Most important features: The leaves are opposite and pinnately divided. The leaflets have a serrated margin. The flowers are large and yellow with red stripes in the flower tube.
Growth structure: The trumpet flower grows as a bush or a small tree. It rarely grows more than 33 feet (10 m) high. The branches are drooping.
Leaves: The leaves are up to 16 inches (40 cm) long with three to eleven tapering, pointed, small leaves. The terminal leaf is 1.6 to 8 inches (4 to 20 cm) long and .4 to 2.4 inches (1 to 6 cm) wide. The ones on the sides are smaller.
Flowers: These grow in simple racemes at the ends of the branches. They have a small, dentate calyx and a corolla tube that is up to 2.4 inches (6 cm) long. At the bottom, the tube is narrow; then it widens, thimble-like, with a limb of five wide lobes. The tube has two long and two short stamens.
Fruits: The fruits are tapering and spindle-shaped. In cross-section, they appear spherical. They are up to 8 inches (20 cm) long and .4 inch (1 cm) thick. The two flaps break open when the fruit is ripe. Inside are many wide-winged seeds about 1 inch (2.5 cm) wide.
Occurrence: Trumpet flowers grow in all tropical and subtropical areas. Originally, they spread from the southern United States to Argentina.
Other names: English "Yellow Elder," "Yellow Bells"; German "Gelber Trompetenstrauch," "Gelbe Tecome"; French "Chevalier"; Spanish "Flor de San Pedro," "Fresnillo," "Gloria," "Roble Amarillo," "Sauco Amarillo," "Trompetilla," "Tronadora."
Additional information: The trumpet flower is among the most popular ornamental plants from the tropics, and it has many folk names. Some *Tabebuia* species (see page 82) have similar flowers and are often given the same names. However, one can easily tell them apart because of their palmately divided or digitate leaves that look like fingers.

Peanut Butter Cassia

Senna didymobotra
Family: *Caesalpiniaceae*, Carob plants

Most important features: The leaves are large and pinnately divided. The yellow flowers grow in erect simple racemes. The buds and bracts are brown.
Growth structure: The peanut butter cassia grows up to 13 feet (4 m) high as an erect bush; it also grows up to 30 feet (9 m) tall as a small tree. The young branches are hairy.
Leaves: The leaves are alternating and 4 to 15 inches (10 to 35 cm) long. The stipules are up to 1 inch (2.5 cm) long. The eight to twenty-two pairs of leaflets are elliptical to tapering and 1.2 to 2.6 inches (3 to 6.5 cm) long.
Flowers: The flowers grow in simple racemes up to 16 inches (40 cm) long. The flowers themselves are 1.2 to 2 inches (3 to 5 cm) long with five golden yellow petals. Two of the ten stamens are larger and horn-shaped. They point to the outside. The ovary is hook-shaped and bent upward.
Fruits: These are up to 4.8 inches (12 cm) long and 1 inch (2.5 cm) wide. They are flattened and dark brown with soft hair. They are cross-ribbed, have no wings, and open lengthwise when ripe.
Occurrence: Widely spread as an ornamental bush as far north as the Mediterranean area, the peanut butter cassia was originally from tropical East Africa.
Other names: English "Popcorn Cassia"; German "Erdnusskassie."
Additional information: The name comes from the unpleasant smell the leaves have. Some people identify this odor with rancid peanut butter; others find it similar to popcorn or even to mice. Sometimes, the plant is confused with *S. Alata*. However, the latter has yellow buds and bracts and winged fruits. Both species were once listed as part of the *Cassia* genus (see page 68).

Kolomona

Senna surattensis
Family: *Caesalpiniaceae*, Carob plants

Most important features: The leaves are pinnately divided. The short, yellow flowers grow almost horizontally in simple racemes close to the ends of the branches. The ovary is hook-shaped and bent upward.
Growth structure: Kolomona grows up to 10 feet (3 m) high as a wide, sweeping bush with thin, drooping branches. As a small tree, it rarely reaches 23 feet (7 m).
Leaves: The leaves are alternating and up to 8.8 inches (22 cm) long. The six to nine pairs of ovate to tapering leaflets are .8 to 1.6 inch (2 to 4 cm) long and approximately .6 inch (1.5 cm) wide.
Flowers: Up to fifteen flowers grow in a simple raceme. The flowers are 1 to 1.6 inch (2.5 to 4 cm) long with five petals on short petioles. Two of the ten stamens are larger and hanging.
Fruits: These grow up to 8 inches (20 cm) long and .6 inch (1.5 cm) wide. They are flat, bare, and bean-shaped. When they are ripe, they are brown and paperlike. They open lengthwise to scatter many flat seeds.
Occurrence: Originally, kolomona spread from India to Polynesia and the northern part of Australia. They are also now planted in the Western Hemisphere, where they sometimes grow wild.
Other names: English "Scrambled Eggs"; German "Rührei-Kassie."
Additional information: The kolomona can withstand poor soil. It needs little care and flowers almost all the time. For this reason, the kolomona and its close relatives are often planted in public gardens. One of its nicknames, scrambled eggs, refers to the dense clumps of yellow flowers that resemble scrambled eggs. People in Southeast Asia use the flowers as a laxative. The leaves and fruits of two closely related species, *S. alexandrina* and *S. angustifolia* are used for the same purpose. All of the species were considered part of the *Cassia* genus at one time.

Glory Bush

Tibouchina urvilleana
Family: *Melastomataceae*, Tibouchina plants

Most important features: The leaves are opposite and have soft hair. The large, violet flowers have five long and five short, clawlike, bent stamens.
Growth structure: The glory bush grows up to 26 feet (8 m) high with four-edged, silky, hairy branches.
Leaves: The leaves are ovate to elliptical, 1.6 to 4.8 inches (4 to 12 cm) long, and .8 to 2 inches (2 to 5 cm) wide. They have silky hair and are delicately dentate on the margins with five main veins. On the upper side, the veins are slightly immersed; on the bottom side, they are firmly emerging. The bottom side of the leaf is denser. It may be lighter in color, or it may be reddish and hairy.
Flowers: The flowers grow individually or a few together at the tips of branches. They are enclosed in reddish bracts that are ovate and .4 to .6 inch (1 to 1.5 cm) long. Flowers are 2.8 to 4.8 inches (7 to 12 cm) long with five petals. The stamens have light appendages. The pistil is up to 1.2 inch (3 cm) long, hook-shaped, and bent upward.
Fruits: The fruits are ovate to tapering capsules. They are .4 to .6 inch (1 to 1.5 cm) long, hairy, and crowned by the remainders of the calyx.
Occurrence: This is a popular ornamental bush that also grows wild in the subtropics. It originated in Brazil.
Other names: English "Princess Flower"; German "Tibouchina."
Additional information: The glory bush grows rapidly and tends to drive out other plants. For this reason, it must be kept under control, even in Hawaii. In the central tropics, it is an evergreen; in cooler or drier areas, it is deciduous, shedding its leaves. In this form, its leaves turn red before falling to the ground. In moderate areas where the summers are warm enough, it can even be kept as a kind of herbaceous plant that freezes back to the ground in winter. Of the five thousand members of this family, approximately 250 *Tibouchina* species are planted as ornamental bushes (see pages 138 and 140).

Forget-Me-Not Tree

Duranta erecta
Family: *Verbenaceae*, Vervain plants

Most important features: The four-edged branches hang. The leaves are opposite. The flowers are small, almost regular violet flowers. The fruits are orange yellow.
Growth structure: As a bush, it grows up to 20 feet (6 m) high, often with thorns up to 1.2 inches (3 cm) long. It rarely grows as a small tree.
Leaves: The leaves are elliptical to obovate, 1 to 2.8 inches (2.5 to 7 cm) long, and .4 to 1.4 inch (1 to 3.5 cm) wide. In the upper half, they often have a serrated margin.
Flowers: These hang in simple racemes up to 6 inches (15 cm) long. The flowers are about .4 inch (1 cm) long with a five-dentate calyx. They have a narrow, crooked corolla tube and five light violet corolla cusps. Often, only the two at the bottom have a darker stripe. Occasionally, they are completely white.
Fruits: These are spherical to ovate stone fruits. They are about .4 inch (1 cm) long and consist of five indistinct parts with a short cusp growing out of the remains of the calyx.
Occurrence: The plant is found in gardens and grows wild in many areas. It was originally from the Western Hemisphere.
Other names: English "Golden Dewdrop," "Pidgeonberry," "Skyflower"; German "Durante"; Spanish "Adonis Morado," "Celosa Cimmarona," "Cuenta de Oro," "Espina de Paloma," "Fruta de Iguana."
Additional information: The forget-me-not-tree has spread from the southern portion of the United States to the southern part of Brazil and to other continents. It is highly poisonous and never eaten raw. Its name is based onthe resemblance of its flowers to the real forget-me-not. Although neither its flowers nor its singular fruits are particularly spectacular, both are present in large numbers almost all year round. This makes the bush attractive in gardens.

Java Glorybower

Clerodendrum speciosissimum
Family: *Lamiaceae*, Labiate

Most important features: The silky leaves are opposite and hairy. The flowers and their petioles are bright red with a long corolla tube and five, curved corolla cusps that point upward and sideward.
Growth structure: This bush grows up to 13 feet (4 m) high with thick branches that have soft hair and that are slightly four-squared.
Leaves: The leaves, ovate to heart-shaped, are 4 to 15 inches (10 to 35 cm) long and 3.2 to 10.4 inches (8 to 26 cm) wide with slightly dentate margins. The veins are immersed on the upper side and emerging on the bottom side.
Flowers: These grow in richly flowering, erect panicles up to 18 inches (45 cm) long. The corolla tube is 1 to 1.6 inch (2.5 to 4 cm) long, and the cusp is .6 to 1 inch (1.5 to 2.5 cm) long. The four stamens and the pistil stand 1.6 to 2 inches (4 to 5 cm) out of the tube and are often bent toward the ground.
Fruits: The fruits are flattened and spherical. They are about .4 inch (1 cm) thick and dark blue. The fruit has three or four parts above the pit. The fruits appear at the slightly thickened petiole; the sepals remain on the fruit.
Occurrence: This ornamental bush is widely spread. It is often grown in the shade. It originated in Java.
Other names: German "Losstrauch," "Pagodenstrauch."
Additional information: The Java glorybower differs from the very similar *C. paniculatum*, which has smaller flowers and partly lobed leaves. Many other of the four hundred *Cleropendrum* species are cultivated (see page 192). The flowers exhibit a large variety of colors and shapes. *C. myricoides*, from tropical Africa, has blue flowers and five corolla cusps. Four of these are lighter and point sideward. The fifth is darker and points toward the ground. The pistil and stamens are bent out of the flower. The harlequin glorybower (*C. trichotomum*) has white flowers in reddish calyxes. This bush is winter hardy.

Coral Plant

Russelia equisetiformis
Family: *Scrophulariaceae*

Most important features: The thin, green, drooping branches have tiny, often barely recognizable leaves. The flowers are tubular and bright red.
Growth structure: The dense coral plant grows up to 3 feet (1 m) high with squared branches.
Leaves: These grow in whorls of three to six on the stronger branches. On thin branches, they grow opposite. The leaves are shaped like a needle and are rarely more than .6 inch (1.5 cm) long.
Flowers: The drooping flowers grow individually or in pairs in the axils of small scales. The calyx grows up to .2 inch (5 mm) long. The corolla tube is approximately .6 to 1 inch (1.5 to 2.5 cm) long with five cusps. The two flowers at the top are connected to one common lobe. Four stamens and one pistil are in the tube.
Fruits: The fruit are spherical capsules. They are .2 (5 mm) long. The pistil is about .6 inch (15 mm) long. Inside are many small brown seeds.
Occurrence: Frequently grown as an ornamental tree, the coral plant probably originated in Mexico.
Other names: English "Fire Cracker Plant," "Fountain Bush"; German "Russelie"; Spanish "Arete de la Cocinera," "Coralillo," "Lluvia de Fuego."
Additional information: Like many cultivated plants, the coral plant is so widely spread today that its place of origin can hardly be determined. However, since its closest relatives come from Mexico, scientists assume that the coral plant also comes from Mexico. As a plant of sunny, somewhat dry locations, it often has no leaves in order to reduce its water loss. The vital functions of the leaves are taken over by green branches. During the more humid season, the other approximately fifty *Russelia* species have flat deciduous leaves. Hummingbirds, attracted by its red flower tubes, often pollinate these plants.

Pomegranate

Punica granatum
Family: *Punicaceae*, Punic apple

Most important features: The five to eight flowers grow with a red, leathery, bell-shaped calyx. The petals are orange red, and there are many stamens. The fruit is spherical and crowned by sepals.
Growth structure: The pomegranate is richly branched and grows up to 20 feet (6 m) high. The bush often has thorns.
Leaves: The shiny leaves are light green, opposite, and tapering to elliptical. They are .4 to 4 inches (1 to 10 cm) long and .2 to 1 inch (0.5 to 2.5 cm) wide.
Flowers: These grow individually or in groups. The calyx is .8 to 1.2 inch (2 to 3 cm) long with triangle cusps. The petals are wrinkly and up to 1.6 inch (4 cm) long. The cusps of the calyx are often as long or longer than the petals.
Fruits: The fruits are yellowish to brownish red and 2.4 to 4.8 inches (6 to 12 cm) long with a leathery exterior peel. Inside are branched septums with a glassy, juicy, wine red hull.
Occurrence: One finds the pomegranate in tropical and subtropical gardens. Originally, it spread from Persia to northwest India.
Other names: German "Granatapfel"; French "Grenadier"; Portuguese "Romanzeiro"; Spanish "Granado."
Additional information: The pomegranate is one of the oldest cultivated plants. It is mentioned in ancient Egyptian descriptions as well as in oriental and Greek sagas, often as a symbol of love and fertility. The wine produced from its juice is considered a love potion. The sweet, juicy seed hull can also be eaten uncooked and can be made into jelly or syrup. The peel, rich in tanning substances, was once used to produce ink and colors for carpets. The genus name goes back to the Roman description, *malus punica*, or Phoenician apple. Some specimens have white or filled flowers. There is also a dwarf specimen, nana.

Burning Love

Ixora coccinea
Family: *Rubiaceae*

Most important features: Leaves are opposite. The flowers grow in false umbels in groups of four or five and are bright red. They have a thin corolla tube and acute corolla lobes.
Growth structure: The burning love is a dense bush that is seldom more than 5 feet (1.5 m) high, but it can reach up to 20 feet (6 m).
Leaves: The leaves are oval to tapering. They are 2 to 4 inches (5 to 10 cm) long and .8 to 2 inches (2 to 5 cm) wide. The leaves are shiny and pointed. At the bottom, they are spherical to heart-shaped and almost sessile.
Flowers: Many flowers grow together in dense inflorescences that are umbrella-shaped. The buds are pointed. The corolla tube is 1 to 1.8 inch (2.5 to 4.5 cm) long. The corolla lobes are widened between the stamens.
Fruits: These are the size of peas. They are dark red, stone fruits crowned by four sepals.
Occurrence: Frequently found in gardens, they are also used as hedges. They were originally from India.
Other names: English "Jungle Flame," "Flame Flower," "Flame of the Wood"; German "Scharlachrote Ixora"; Spanish "Cruz de Malta," "Santa Rita."
Additional information: The flowers of this bush are often scarlet red; however, some species have orange to yellow flowers. Because the inflorescences keep well, even when cut, they are often part of Buddhist flower offerings. Other members of the approximately three hundred *Ixora* species are also cultivated. *I. chinensis* and *I. javanica* are very similar. They are easy to recognize because of their round corolla lobes, petioles, and less dense inflorescences. *I. casei* has larger, simple umbels that are crimson and almost spherical. *I. finlaysoniana* is a high bush with small, white flowers. *I. Odorata*, from Madagascar, has white flowers with a reddish-streaked corolla tube that is 4.8 inches (12 cm) long. It has an attractive aroma.

Pride of Trinidad

Warszewiczia coccinea
Family: *Rubiaceae*

Most important features: The large leaves are opposite. The inflorescences have long axils, small yellow flowers, and bright red leaves.
Growth structure: The pride of Trinidad grows up to 50 feet (15 m) high with firm, squared branches. When cultivated, it is often kept as a bush.
Leaves: The leaves are elliptical to tapering. They are 6 to 24 inches (15 to 60 cm) long and 2.4 to 12 inches (6 to 30 cm) wide with fifteen to sixty steeply branching lateral veins. Between the veins, the leaves are often wavy. Between the petioles are ovate to triangular stipules that are up to 1.2 inch (3 cm) long.
Flowers: The drooping inflorescences are up to 32 inches (80 cm) long. On the upper side, they are dense with groups of flowers .2 inch (5 mm) long. Five flowers grow from one petiole. The calyx is often less than .2 inch (5 mm) long. It is green and has one or two flowers per group. One sepal is enlarged, elliptical, and up to 4.8 inches (12 cm) long and 1.6 inch (4 cm) wide.
Fruits: The fruits are almost spherical and about .2 inch (0.5 cm) long. They have two flaps with many tiny seeds.
Occurrence: Occasionally, one sees them in gardens. They originated in the Western Hemisphere.
Other names: English "Chaconia," "Wild Poinsettia"; German "Stolz von Trinidad"; Spanish "Barba Gallo," "Crucero," "Guna," "Sangrenaria."
Additional information: The pride of Trinidad is one of the national symbols of Trinidad and Tobago. It is named after a recent Spanish governor, José Maria Chaón and called Chaconia. It always flowers around the country's Independence Day, August 31st. The large, red leaves attract hummingbirds that drink the nectar out of the flowers.

Ashanti Blood

Mussaenda erythrophylla
Family: *Rubiaceae*

Most important features: The pink or red leaves are set opposite. At least one sepal is enlarged and grows up to 4 inches (10 cm) long.
Growth structure: In cultivation, it is often an erect bush and seldom grows more than 10 feet (3 m) high. As a wild plant, it climbs up to 30 feet (9 m) high. The young branches have red hair.
Leaves: The tapering leaves are heart-shaped to ovate. They are 2.4 to 6 inches (6 to 15 cm) long and 1.2 to 3.6 inches (3 to 9 cm) wide.
Flowers: The flowers grow on dense or umbrella-shaped panicles with five flowers on each panicle. They have a narrow, .6 to 1 inch (1.5 to 2.5 cm) long tube that is reddish and hairy around the tube. The limb of the corolla is light yellow to white and .3 to .8 inch (8 to 20 mm) wide.
Fruits: The hairy fruits are elliptical to tapering. They are about .8 inch (2 cm) long and .4 inch (1 cm) thick.
Occurrence: One finds the Ashanti blood in gardens and parks. It was originally from West Africa.
Other names: English "Red Flag Bush"; German "Rotblättrige Mussaenda."
Additional information: In the wild specimen of this species, only a few of the sepals and outer flowers of each inflorescence have the typical bright scarlet to crimson color. The leaves are similar to deciduous leaves in size, shape, and texture. With cultivated specimens, most of the sepals of all the flowers are enlarged and weaker in color. This is especially true in Asia with the popular Queen Sirikit. It often only produces a few yellow flowers between masses of pink leaves. The wild specimen of the virgin tree (*M. philippica*) has only one white leaf; the rest of the leaves and the flowers are light yellow. Other members of the approximately one hundred *Mussaenda* species are occasionally cultivated. For example, in Southeast Asia, *M. frondosa* has one white leaf and firm, orange yellow flowers.

Malaysian Orchid Tree

Medinilla magnifica
Family: *Melastomataceae*

Most important features: The branches are four-squared. The flowers grow in drooping panicles up to 20 inches (50 cm) long with large pink bracts.
Growth structure: This wide bush grows up to 10 feet (3 m) high. The branches are hairy at the knots.
Leaves: The leaves grow opposite with a heart-shaped bottom. They are sessile, oval, and 8 to 12 inches (20 to 30 cm) long. The leaves are pointed and leathery with strongly emerging veins on the bottom side.
Flowers: The flowers are up to 1 inch (2.5 cm) long. They are pink red or light violet with a cup-shaped calyx, five petals, and ten stamens. The stamens each have two little knots and one long, crooked, pale lilac anther. The bracts are ovate and up to 8 inches (20 cm) long. At the beginning, the bracts encompass the entire panicle; later the panicles stand out.
Fruits: The fruits are violet berries up to .4 inch (1 cm) in diameter. They are crowned by a ring-shaped remnant of the calyx.
Occurrence: The Malaysian orchid tree is most frequently seen as a potted plant in front of a house. Originally, it was from the Philippines.
Other names: English "Rose Grape"; German "Medinilla."
Additional information: The Malaysian orchid tree is used as an outdoor ornamental plant. When brought indoors, it often suffers from lack of light and too little humidity. Despite its size, it prefers growing in the forks of the branches of large trees, where it is hard to see. Even in the tropics, the tree is seen more often in a pot than growing in the garden. One can find others of the approximately four hundred *Medinilla* species growing on the ground. Most of those do not have the same striking bracts as the Malaysian orchid tree. In other respects, however, they are very similar, including the richer flowering *M. speciosa* in the mountainous regions of Malaysia and Indonesia. Other members of *Melastomataceae* grow almost everywhere in the tropics (see also pages 130 and 140).

Barbados Cherry

Malpighia emarginata
Family: *Malpighiaceae*

Most important features: The flowers have five petals. The sepals have two oil glands. The petals grow on petioles and are pink to red. The petal on top is a little longer with a frayed margin.
Growth structure: This widely variable, low bush grows up to 20 feet (6 m) high like a little tree.
Leaves: The leaves are opposite and ovate to tapering. They are .8 to 3.2 inches (2 to 8 cm) long and .4 to 2 inches (1 to 5 cm) wide. The leaves are a shiny dark green.
Flowers: The flowers grow two to five in small simple umbels. They are .6 inch (15 mm) long with ten yellow stamens.
Fruits: The fruits are apple-shaped to almost spherical. They are .4 to 1.2 inch (1 to 3 cm) long and red. Often, they have some longitudinal grooves.
Occurrence: One finds the Barbados cherry in the Western Hemisphere and on Hawaii. Occasionally, they are cultivated in Southeast Asia. Originally, they spread from the Caribbean area and the northern part of South America.
Other names: English "West Indian Cherry"; German "Barbadoskirsche"; French "Cerise de Cayenne," "Cerise de Antilles"; Portuguese "Cerejeira das Antilhas"; Spanish "Acerola," "Escobillo," "Grosella," "Semeruco."
Additional information: Children love the taste of the Barbados cherry, although some of the wilder specimens are quite sour. Because of its high concentration of vitamin C, sweeter varieties were cultivated and grown on large plantations. Unfortunately, most efforts to grow this small, aromatic fruit have been abandoned because the fruits never became popular enough and because synthetic vitamin C is much cheaper to produce. As an ornamental plant, one sees the closely related *M. coccigera*. It is often a smaller bush, seldom reaching 7 feet (2 m) with small leaves that grow up to .8 inch (2 cm) long. Many pointed thorns grow on the edge of the leaves.

Singapore Rhododendron

Melastoma malabathricum
Family: *Melastomataceae*

Most important features: The leaves are set opposite with three to seven main veins. The flowers are light red violet and occasionally white. They have ten stamens; five are erect and entirely yellow, and five have long, violet anthers bent in the shape of a claw.
Growth structure: The Singapore rhododendron grows up to 10 feet (3 m) high as a bush with four-squared branches. It may also grow up to 16 feet (5 m) tall as a tree.
Leaves: The leaves are elliptical to tapering ovate. They grow 2.4 to 6 inches (6 to 15 cm) long and .8 to 2.6 inches (2 to 6.5 cm) wide. The leaves are hairy, and the lateral veins are almost shootlike between the main veins.
Flowers: These grow in little flowering panicles at the ends of the branches. They are 1.4 to 2.8 inches (3.5 to 7 cm) long. The area under the flower is scaly. The flowers have five to eight petals. The pistil is as long as the long stamens.
Fruits: These are urn-shaped, fleshy, and about .4 inch (1 cm) long. They are crowned by the remains of the sepals. Inside, the flesh is blue black, and the seeds are orange.
Occurrence: This is a very popular bush in Southeast Asia, especially on fallow land and along roadsides. Originally, the Singapore rhododendron spread from Mauritius to Polynesia and from the Himalayas to Australia.
Other names: English "Straits Rhododendron"; German "Singapur-Rhododendron."
Additional information: The Singapore rhododendron is not related to the real *Rhododendron*, which has more than eight hundred species in the Asiatic mountains. Its fruits are edible, but they are tasteless. Yet children like to eat them because they stain the mouth and tongue dark violet to almost black. The genus and the entire family owe their name to this characteristic. In Greek, *melas* means black and *stoma* means mouth.

Pineapple Guava

Acca sellowiana
Family: *Myrtaceae*, Myrtle

Most important features: The leaves on the bottom side are white to gray and feltlike. The flowers look like a brush made of numerous crimson filaments in the middle. The flower has four petals that curl downward. On the upper side, the leaves are crimson.
Growth structure: The pineapple guava is a bush or small tree that grows up to 20 feet (6 m) high.
Leaves: The leaves are opposite, elliptical, and up to about 2.8 inches (7 cm) long. On the upper side, they are dark green. On the midrib, they are densely hairy. Typically, the rest of the leaf is completely bare.
Flowers: These grow individually or in a small group. They are 1.2 to 1.6 inch (3 to 4 cm) in diameter with four sepals. The upper side is red brown and the lower side is feltlike. Four petals alternate with the sepals (see photo at right). The stamens carry small anthers that change color. Before they unfold, they are light red; then, they are yellow with pollen. Finally, when the pollen is gone, they turn dark red before they are shed. The pistil, which rises up in the middle, is slightly thicker and longer than the filaments, but otherwise it is quite similar.
Fruits: The fruits are berrylike, slightly tapering, and 2 to 2.8 inches (5 to 7 cm) long. When they are entirely ripe, they are red brown. The fruits are crowned by sepals and have whitish flesh and numerous seeds.
Occurrence: Frequently cultivated, the pineapple guava was originally from the southern part of South America.
Other names: Portuguese (Brazil) "Feijoa"; German "Brasilianische Guave"; Spanish "Guayabo del Brasil."
Additional information: The aroma is similar to that of a pineapple. The sour flesh is usually eaten raw. In South America, it is made into a jam.

Cherry Pie

Lantana camara
Family: *Verbenaceae*, Vervain plants

Most important features: The branches are four-squared. The leaves are set opposite and are slightly wrinkled. Often the flowers are yellow in the center of the inflorescences and orange, red, or pink away from the center.
Growth structure: Cherry pie grows up to 10 feet (3 m) high as a subshrub. It is only slightly woody and often has curved thorns.
Leaves: The leaves are ovate, 1.6 to 4 inches (4 to10 cm) long, and hairy with a serrated margin. When compressed, the leaves exude an unpleasant odor.
Flowers: The inflorescences are 1.2 to 2 inches (3 to 5 cm) wide. The flowers on top have narrow tubes and five spread lobes. The outer flowers are .3 inch (8 mm) in diameter, with a tube up to .8 inch (2 cm) long and irregular lobes. The inner flowers are smaller and more regular in shape.
Fruits: These are black, spherical, and about .2 to .3 inch (5 to 8 mm) long.
Occurrence: One finds cherry pie in almost all tropical and subtropical areas. Originally, it grew from Texas to South America.
Other names: The species has hundreds of folk names, among others English "Shrub Verbena," "Wild Sage"; German "Wandelröschen"; Spanish "Bandera Española," "Camará," "Cinco Negritos," "Coronitas del Sol."
Additional information: The flowers of the cherry pie attract butterflies, but this plant is so poisonous that hardly any other animal dares to touch it. It has been introduced as an ornamental plant in many places, but it has turned out to be an aggressive weed. This is why it is also referred to as the "curse of India." In many areas, it is only kept in check by importing natural parasites. Today, less aggressive specimens are available. They have been crossbred with the colors of other flowers. The closely related shrub, verbena (*L. Montevidensis*), is often planted as a ground cover. Verbena has light violet flowers.

Bloodflower

Asclepias curassavica
Family: *Asclepiadaceae*, Silkweed

Most important features: The bloodflower has large amounts of milky sap in all of its parts. The leaves are set opposite, and the flowers are red and yellow orange.
Growth structure: The plant grows up to 4 feet (1.2 m) high. The branches are slightly woody.
Leaves: The leaves are 3.2 to 6 inches (8 to 15 cm) long and .4 to .8 inch (1 to 2 cm) wide. They taper at both tips. On the upper side, they are dark green; on the bottom side, they are often slightly blue.
Flowers: The flowers grow in simple umbels at the ends of the branches. The flowers are up to .4 inch (1 cm) long. The petals are light red and often curled downward. Above them is a yellow to orange corona and five bag-shaped segments, each bent toward the middle.
Fruits: The fruits are narrow and cigar-shaped. They grow in pairs, each up to 6 inches (15 cm) long. The seeds are white with a silky tuft of hair.
Occurrence: Bloodflower is widely spread as an ornamental plant and as a weed. It originated in the Western Hemisphere.
Other names: English "Bastard Ipecacuanha," "Butterfly Weed," "Indian Root," "Orange Milkweed"; German "Seidenpflanze"; French "Ipéca Sauvage"; Portuguese "Oficial da Sala"; Spanish "Algodoncillo," "Corcalito," "Flor de Sangre," "Mata Caballo," "Yuquillo."
Additional information: The hair of the seeds is too smooth and fragile to be useful. The plant is feared as a weed because it spreads very quickly. It is so poisonous that it is rarely eaten. Only the grubs of one insect (*Danaus chrysippus*) eat the leaves. Then, they become poisonous themselves. Butterflies often visit the flowers. In order to reach the nectar, they stick their beaks in the coronas, get stuck, and transfer small packets of pollen.

Brazilian Plume Flower

Justicia carnea
Family: *Acanthaceae*, Acanthus

Most important features: The stem is four-squared. The flowers grow in dense panicles. They are often pink with two lips.
Growth structure: The plant grows up to 7 feet (2 m) tall as a firm subshrub. It is only slightly woody.
Leaves: The leaves are opposite and tapering at both tips. They grow up to 10 inches (25 cm) long and 4.8 inches (12 cm) wide. They are slightly wavy on the margins, and they are often a little vaulted between the veins.
Flowers: These grow in panicles up to 8 inches (20 cm) long at the ends of the branches. The calyx has five small lobes about .4 inch (1 cm) long. The corolla is up to 2.8 inches (7 cm) long with an almost erect tube that is curled downward. The labium has three lobes and is erect. The two stamens are below the two lower lobes, which are slightly bent.
Fruits: The fruits are tapering, slightly club-shaped capsules with one or two disc-shaped seeds per compartment. The fruits are seldom seen.
Occurrence: The Brazilian plume flower grows in gardens and parks. It prefers shade. Originally, it was from Brazil.
Other names: English "Flamingo Plant"; German "Jacobinie"; Spanish "Isopo Rojizo," "Tango Rojizo."
Additional information: The Brazilian plume flower can be a firm, herbaceous plant or a juicy bush. Frequently propagated by cuttings, it often does not develop any fruit. Specimens with lilac, red, or white flowers are rare. Previously, the plant was called *Jacobinia carnea* or *Cyrtanthera magnifica*. However, today it is categorized as a member of the *Justicia* genus (see page 216). *J. rizzinii* (see photo at right) is a decorative dwarf bush that is also sold as *Jacobinia pauciflora*. The dried leaves of *Justicia pectoralis* are sniffed as an hallucinogenic drug.

Sanchezia

Sanchezia nobilis
Family: *Acanthaceae*, Acanthus

Most important features: The leaves are opposite. The midrib and the lateral veins are light yellow to white. The flowers are yellow and tubular. They grow in the axil of the red brown to red bracts.
Growth structure: This is a wide, slightly woody bush that grows up to 7 feet (2 m) high. The branches are square-shaped and fragile. They are thickened and often reddish.
Leaves: The leaves are ovate to tapering and elliptical. They are 4 to 12 inches (10 to 30 cm) long and 2 to 4.8 inches (5 to 12 cm) wide. The margins are dentate.
Flowers: The flowers grow in groups of up to ten erect spikes. These are 2 to 16 inches (5 to 40 cm) long. All the flowers point to one side. The bracts are ovate and 1 to 2.2 inches (2.5 to 5.5 cm) long. The corolla tube is erect and points steeply upward. The tube is 1.8 to 2.2 inches (4.5 to 5.5 cm) long with five small, curled cusps. The two stamens and the pistil extend out of the tube.
Fruits: The fruits are tapering capsules with two compartments holding six to eight seeds.
Occurrence: Sanchezia grows in parks and gardens. It was originally from Ecuador.
Other names: Interestingly, the plant is called "Sanchezia" everywhere.
Additional information: The sanchezias are often planted for their decorative leaves. *S. parvibracteata* has rather insignificant bracts in its inflorescences. The wild species, *S. Nobilis*, can rapidly overgrow an entire garden. *S. glaucophylla* is planted more often. It has lighter stripes that are very wide above the leaf veins. One can find similar leaves with some specimens of *Aphelandra speciosa* from the same family. However, these are smaller and have yellow flowers that stand out, cross-shaped, in four directions at the bracts of the inflorescences.

Bowstring Hemp

Calotropis gigantea
Family: *Apocynaceae*, Dogbane

Most important features: All parts of the plant have a milky sap. The leaves are set opposite. They are stiff, leathery, and a waxy bluish gray. The flowers are cream white to light violet. They grow in groups that form a sort of pyramid.
Growth structure: The bush has firm branches and grows from 3 to 16 feet (1 to 5 m) high. Occasionally, it grows as a small tree.
Leaves: The leaves have short petioles and a heart-shaped bottom. They are elliptical, 3.2 to 8 inches (8 to 20 cm) long, and 1.6 to 4.8 inches (4 to 12 cm) wide with pointed ends.
Flowers: The inflorescences appear like false umbels. They are 1 to 1.8 inch (2.5 to 4.5 cm) long. The waxy petals widen. The pyramid developed by the corona is about .6 inch (1.5 cm) high and often slightly darker than the petals. The stigma is greenish yellow.
Fruits: These are bluish green, 2.8 to 4 inches (7 to 10 cm) long, and 1 to 1.6 inch (2.5 to 4 cm) thick. The fruits are bent at the bottom and have many seeds. Each seed has a hairy tuft.
Occurrence: One finds the bowstring hemp on coasts and in relatively dry locations. It originated in Pakistan and Southeast Asia.
Other names: English "Crown Flower," "Giant Milkweed," "Ivory Plant," "Madar"; German "Kronenblume."
Additional information: The petioles of the bowstring hemp supply tough fibers, which can even be even used as fishing lines. The poisonous, milky sap has many medicinal applications. When dry, it turns into a rubber-like substance. The flowers stay fresh for a long time and are popular in flower arrangements and floral wreaths. The closely related rooster tree (*C. procera*) is also cultivated. Its leaves are larger. They grow up to 12 inches (30 cm) long; yet, the flowers are slightly smaller. They are only about .6 to 1 inch (1.5 to 2.5 cm) long. The petals are bell-shaped and drooping. They often have a distinct, dark violet spot at the top.

Guava

Psidium guajava
Family: *Myrtaceae*, Myrtle

Most important features: The leaves are set opposite. The lateral veins are steeply branching. The white flowers have four to five sepals and petals and many stamens. The fruits are lemon-shaped and crowned by sepals.
Growth structure: Guava grows as a wide bush with delicately hairy, four-squared branches. Occasionally, it reaches 33 feet (10 m) as a tree. The older branches shed their bark in strips.
Leaves: The gray green leaves are elliptical to tapering. At the tips, they are often spherical. They are 1.6 to 5.6 inches (4 to 14 cm) long and 1.2 to 2.4 inches (3 to 6 cm) wide with twelve to twenty pairs of lateral veins that are immersed on the upper side and that emerge on the bottom side.
Flowers: The flowers grow individually or in groups of up to three. They are 1 to 1.6 inch (2.5 to 4 cm) long with an inferior ovary.
Fruits: The fruits are spherical to pear-shaped. When they are ripe, they are yellow or pink and 1.2 to 4 inches (3 to 10 cm) long with white or pink flesh and many little seeds.
Occurrence: Guava is planted everywhere in the tropics and often grows wild. Originally, it was from the Western Hemisphere.
Other names: German "Guave"; French "Goyavier"; Portuguese "Goiaba"; Spanish "Guayabo."
Additional information: Guavas are rich in vitamin C, iron, and other nutrients. They are often made into juices, jellies, or compotes. When raw, they have very little taste and a musky smell. However, some other members of the species are pleasant smelling and seedless. The wood of the guava is firm and long lasting, but the trunks are often too thin to be used. *P. cattleianum* is often cultivated. It has dark red fruits. Its leaves are pointed at the bottom with barely emerging veins.

Cape Jasmine

Gardenia augusta
Family: *Rubiaceae*

Most important features: The leaves are set opposite. They are dark green and shiny. Occasionally, they are yellow and white. The flowers are white and often shaped like a rose.
Growth structure: This is a dense, bushy plant that grows up to 7 feet (2 m) high. It may grow up to 40 feet (12 m) tall in the wild.
Leaves: The leaves are ovate to tapering. They are 2 to 6 inches (5 to 15 cm) long and .8 to 2.8 inches (2 to 7 cm) wide with tiny triangular stipules between the two petioles.
Flowers: The flowers grow individually. They are 2.4 to 3.2 inches (6 to 8 cm) wide. The flowers are white, then yellow when wilting, and finally dark in the middle. The flowers have an inferior ovary, long slender calyx cusps, and a narrow corolla tube that is about 1.2 inch (3 cm) long. The species that grow wild have five to seven corolla cusps.
Fruits: The fleshy fruits are elliptical to ovate and .6 to 1.2 inch (1.5 to 3 cm) long. They are orange and crowned by five calyx tips. A longitudinal rib runs down each tip.
Occurrence: One finds the cape jasmine in gardens. It originated in Southeast Asia and the southern part of Japan.
Other names: English "Common Gardenia"; German "Gardenie."
Additional information: The cape jasmine is a very popular ornamental plant, often sold by florists. However, it is, sensitive and easily sheds its buds. Because the flowers are not very durable, they are not sold as cut flowers. Some specimens have two corollas. Like the Arabian jasmine (*Jasminum sambac*, see page 192), the strong smelling flowers are used to give aroma to jasmine tea. Some of the other approximately sixty *Gardenia* species are also cultivated, especially the sensitive, little South African *G. Thunbergia*. It has a corolla tube that grows up to 2.8 inches (7 cm) long.

Coffee

Coffea arabica
Family: *Rubiaceae*

Most important features: The leaves are set opposite and have a wavy margin. The flowers are white and grow in bunches in the axils. The fruits are small and red and hang on short petioles.
Growth structure: This is a small tree that grows up to 20 feet (6 m) high with horizontal branches. When cultivated, it is often kept as a bush.
Leaves: The leaves are elliptical to tapering, dark green, and shiny. They are 3.2 to 10 inches (8 to 25 cm) long and 1.2 to 4 inches (3 to 10 cm) wide with a tiny, wide, triangular stipule between the petioles.
Flowers: Five to eight flowers grow in bunches. They have a wonderful aroma. The flowers have a short corolla tube with an inferior ovary.
Fruits: The fruits are spherical to elliptical and .6 inch (15 mm) long. They have a membranous hull. Each fruit may have two seeds. These are flattened at the sides that face each other.
Occurrence: Originally from the Ethiopian highlands, this tree is grown in many tropical countries, often under shade trees, at altitudes ranging from 2,000 to 4,000 feet (600 to 1,200 m).
Other names: German "Kaffee"; French, Portuguese, and Spanish "Café"; Spanish also "Cafeto."
Additional information: Coffee is one of the most important cultural plants. Approximately 25 million people earn their living from coffee, and about one-third of the people on earth drink coffee. The coffee beans are the seeds of the coffee berry. The fruity flesh is removed right after the harvest; however, the beans are not roasted until they reach the consumer country. Besides Arabian coffee, which provides almost three-fourths of the worldwide production, two other species are cultivated: Robusta coffee accounts for most of the remaining production. It comes from *C. canephora*, a stronger specimen that grows at lower altitudes. Liberian coffee (*C. liberica*) only represents about one percent of the worldwide market.

Blue Potato Bush

Solanum rantonnetii
Family: *Solanaceae*, Nightshade

Most important features: The leaves are alternating and simple. The flowers are dark violet and almost spherical. They have five recognizable segments. In the middle, the flowers are yellow. The stamen cone is flattened.
Growth structure: This plant grows up to 8 feet (2.5 m) high as a bush with drooping or climbing branches. When cultivated, it is often pruned into a little tree with a spherical top.
Leaves: The leaves are ovate to tapering, herbaceous, and slightly pointed. They are 2.4 to 4 inches (6 to 10 cm) long and often have a wavy margin.
Flowers: The flowers grow individually or in small groups. They have a small green calyx and are .8 to 1.2 inch (2 to 3 cm) wide. Often the corolla is wrinkly. In the middle, the flower has a small, yellow star, rays of which run straight toward the margin and end as a pointed tip.
Fruits: The fruits are red, apple-shaped berries. They are about 1 inch (2.5 cm) long and are rarely seen.
Occurrence: One finds this plant in gardens and in pots on a terrace. Originally, they were from Paraguay and Argentina.
Additonal Names: English "Paraguay Nightshade"; German "Enzianstrauch," "Blauer Kartoffelstrauch."
Additional information: The blue potato bush does not produce potatoes, but like the potato, it belongs to the gigantic *Solanum* genus. This genus includes approximately 1,700 species (see also pages 104 and 198). The shape of the flower is reminiscent of a potato. However, the color resembles climbing nightshade or bittersweet (*S. dulcamara*). As a plant of the southern subtropics, the blue potato bush can stand light frost of up to 27°F (–3°C). Although the branches may freeze, new ones usually sprout from the older wood.

Yesterday-Today-and-Tomorrow

Brunfelsia pauciflora
Family: *Solanaceae*, Nightshade

Most important features: The flowers are violet, but they fade rapidly. They have five petals and a very narrow tube with four stamens inside.
Growth structure: This plant grows up to 10 feet (3 m) high. Specimens grown in cultivation are often not as tall.
Leaves: The leaves are alternating and elliptical to tapering. On the upper side, they are dark green. Different specimens vary widely in size (see below).
Flowers: The flowers are 1 to 4 inches (2.5 to 10 cm) in diameter. The corolla tube is about half as long with a white ring at the opening.
Fruits: The fruits are ovate, slightly fleshy, capsules with two compartments and oval seeds.
Occurrence: This popular ornamental tree was originally from Brazil.
Other name: German "Brunfelsie."
Additional information: Like many nightshades, the yesterday-today-and-tomorrow is poisonous. At one time, some people made intoxicating drugs out of it; and an extract from its root was thought to cure snakebites. It owes its name to the rapid color change of its flowers. Shortly after they unfold, they are dark violet, but they rapidly become lighter. On the third day, at the latest, they turn white. They drop shortly afterward. Two forms are cultivated in gardens. Floribunda seldom grows as tall as a person. Its leaves are only 2.8 inches (7 cm) long and .8 inch (2 cm) wide. It has many small flowers. Macrantha is a larger plant. The leaves grow up to 10 inches (25 cm) long and 2.8 inches (7 cm) wide, and the flowers can be up to 2 to 4 inches (5 to 10 cm) in diameter. *B. americana* has yellowish white flowers with a very long corolla tube that attracts insects at night with a seductively sweet fragrance. This specimen is also called lady of the night.

Cape Plumbago

Plumbago auriculata
Family: *Plumbaginaceae*, Plumbaginaceous plants

Most important features: The flowers have five petals and are light blue with white sides. The corolla tube is long and narrow. The calyx has sticky hair.
Growth structure: This bush has drooping branches that grow up to 7 feet (2 m) long. If supported, the branches may grow twice as long.
Leaves: The leaves are alternating, tapering, and elliptical to obovate. They are 1.2 to 3.6 inches (3 to 9 cm) long and .4 to 1.2 inch (1 to 3 cm) wide. The petiole is shaped like a little ear that is embracing the stem.
Flowers: The flowers grow in simple racemes between .8 and 2.4 inches (2 to 6 cm) long. They grow at the ends of the branches and are almost sessile. The calyx is tubular, about .4 inch (1 cm) long, and often slightly reddish. The corolla tube is 1 to 1.6 inch (2.5 to 4 cm) long. The spread corolla cusps are .4 to .6 inch (1 to 1.5 cm) long, obovate, and rounded at the tip.
Fruits: The fruits grow up to .3 inch (8 mm) long. They are club-shaped capsules. Until they are ripe, they are enclosed in the calyx. Then, the five flaps on the bottom open up.
Occurrence: This plant is found in many tropical and subtropical gardens. Originally, it grew in East Africa.
Other names: English "Leadwort"; German "Kap-Bleiwurz"; Spanish "Azulina," "Celestina," "Jasmín Azul," "No-me-olvides," "Umbela."
Additional information: The cape plumbago can withstand frost to 18°F (–8°C). It is especially beautiful in areas with a clear, dry period because too much rain destroys its look. Extracts of the plant are used in folk medicine; however, it tends to color the skin of the patient lead gray. Other, related species include *P. scandens,* which does not have the same shape petiole but is otherwise quite similar. *P. indica* has pink flowers, and *P. zeylanica* has white or very light violet petals that are more pointed. The juice of all these species is very irritating to the skin.

Good Luck Plant

Cordyline fruticosa
Family: *Dracaenaceae*, Dracaena-like plants

Most important features: The good luck plant is a small-branched bush with ring-shaped leaf scars at the petioles. The leaves accumulate at the ends of the branches. They are tapering. When cultivated (see photo at right), the leaves are often pink, red, or purple in different shades that run lengthwise in stripes.
Growth structure: The plant grows up to 23 feet (7 m) high. The branches are steeply erect. Occasionally, the plant has no branches.
Leaves: The alternating leaves grow from petioles. They are 8 to 24 inches (20 to 60 cm) long and 2 to 4.8 inches (5 to 12 cm) wide with lateral veins. The leaves branch off at very steep angles.
Flowers: The flowers are white to violet and .3 to .6 inch (8 to 15 mm) long. They grow in drooping panicles that are more than 24 inches (60 cm) long. The flowers have spikelike lateral axes that are up to 8 inches (20 cm) long.
Fruits: The fruits are spherical berries about .3 inch (8 mm) long. They are dark red and often have only one seed.
Occurrence: Often found in gardens, the good luck plant was widely spread even before the arrival of Europeans in Asia and Polynesia.
Other names: English "Ti Plant," "Tree-of-Kings"; German "Strauchige Keulenlilie"; Spanish "Caña de Indio," "Palmita Roja."
Additional information: The good luck plant is often placed around houses in Southeast Asia and Polynesia to defend against evil spirits. Because it seldom blooms, the flowers are considered a particularly good omen. In addition to providing good luck, the plant is useful. Its leaves do not shrink when they dry, so they are used in producing mats, receptacles, fans, and skirts. On many Polynesian islands, the roots are a food staple. The roots are cooked, baked, or fermented to create alcoholic beverages. The green specimen of the plant, called ki or ti, was brought to Hawaii by the Polynesians.

Chenille Plant

Acalypha hispida
Family: *Euphorbiaceae*, Wolf's milk

Most important features: The leaves are nettle-shaped. The drooping inflorescences look like long, red tails.
Growth structure: The chenille plant grows as a bush up to 10 feet (3 m) high. The young branches are only slightly woody.
Leaves: The leaves are alternating, tapering, and ovate with serrated margins. They are 4 to 8 inches (10 to 20 cm) long and 2.8 to 6 inches (7 to 15 cm) wide. The leaves are herbaceous with clearly emerging veins on the bottom side.
Flowers: The individual flowers are tiny and unisexual. The male ones are almost never visible; the female ones have thin hair and crimson flower parts. The catkins are 4 to 20 inches (10 to 50 cm) long and .4 to .6 inch (1 to 1.5 cm) thick. They are soft and hang down.
Fruits: The fruits are tiny and spherical. When ripe, they break into three parts.
Occurrence: This is one of the most popular tropical garden plants. Its place of origin is uncertain but may be New Guinea.
Other names: English "Philippine Medusa," "Red Cats-Tail"; German "Katzenschwanz," "Raues Nesselblatt"; French "Queue de Chat."
Additional information: The chenille plant is very popular because its long, red inflorescences are present almost all year long. The soft bloom is created by the pistils of the unisexual flowers. In cultivation, one rarely finds male plants. The female plants are propagated by cuttings. Some inflorescences are used in flower arrangements. Occasionally, one can find specimens of the chenille plant with dark purple or cream white inflorescences. Of more than four hundred *Acalypha* species, only the chenille plant and *A. wiltesiana* (see page 178) are cultivated.

Crimson Bottlebrush

Callistemon citrinus
Family: *Myrtaceae*, Myrtle

Most important features: The plant is slender with leaves that smell like lemons when they are crushed. The inflorescences look like bright red bottle brushes.
Growth structure: This is a bush or small tree that grows up to 40 feet (12 m) high. Some of the branches are partly erect; some of them are partly drooping.
Leaves: The leaves are alternating, very narrow, and stiff. They grow up to 4 inches (10 cm) long. The leaves have many minute holes.
Flowers: The numerous flowers grow in dense inflorescences up to 4.8 inches (12 cm) long. The inflorescences grow at the ends of the branches. The individual flowers are small with five insignificant sepals and petals. The many red filaments have yellow anthers and a yellow green, shining ring of nectar around the pistil.
Fruits: The fruits are urn-shaped and woody. They are up to .4 inch (1 cm) long and crowned by sepals.
Occurrence: One finds the crimson bottlebrush as an ornamental tree in gardens in tropical areas and in areas with mild winters. The plant originated in Australia.
Other names: German "Schönfaden," "Zylinderputzer."
Additional information: The crimson bottlebrush comes from dry areas. Therefore, it has adapted to fires. The oils of its leaves burn explosively. This robs the fire of oxygen so that the branches are only slightly scorched. The fruits often open only after a fire when the seeds will find favorable conditions to germinate, and the young plants will have little competition. Of the approximately thirty *Callistemon* species, the very similar weeping bottlebrush (*C. viminalis*) is also planted. It differs from the former because of its drooping branches and slightly longer inflorescences. One rarely sees the pale yellow *C. pallidus*.

Chinese Hibiscus

Hibiscus rosa-sinensis
Family: *Malvaceae*, Mallow plants

Most important features: The upper halves of the leaves have a serrated margin. The flowers are large and often bright red with five undivided petals and one midrib. The midrib has many yellow anthers in the upper third and five headlike, thickened, velvety stigmas at the tip.
Growth structure: The plant grows erect, up to 20 feet (6 m) high.
Leaves: The leaves are alternating, ovate to oval, tapering, and herbaceous. They are 1.6 to 6 inches (4 to 15 cm) long and 1 to 4 inches (2.5 to 10 cm) wide with delicately hairy stipules.
Flowers: The flowers grow individually. They are often 4 to 6 inches (10 to 15 cm) wide. The slender leaves of the epicalyx bracts are located at the bottom of the calyx.
Fruits: The fruits are long capsules with many seeds. When they are ripe, five flaps open.
Occurrence: The Chinese hibiscus is one of the most popular ornamental bushes of the tropics and subtropics. Originally, it was from Southeast Asia.
Other names: English "Chinese Rose," "Shoe Flower"; German "Chinesischer Roseneibisch"; French "Rose de Chine"; Spanish "Clavel Japonés," "Rosa de China."
Additional information: The Chinese hibiscus has been cultivated in Asia for a very long time. For many years, people used the juice of its flowers to blacken the color of their hair, eyebrows, and even shoes. Besides the wild specimens, many specimens are domesticated for breeding. These have white, speckled, or torn leaves and a pink, orange, yellow, white, or multicolored corolla that rarely grows up to 12 inches (30 cm) high. Many other *Hibiscus* species are also cultivated (see page 170). Within a day, the color of the flowers of the rose (*H. mutabilis,* see photo at right) changes from pale pink to deep red. The leaves of the changeable rose have five lobes and soft hair. The leaves of the rose of Sharon (*H. syriacus*) have three lobes. The flowers are violet or occasionally white. This plant is winter hardy.

Coral Hibiscus

Hibiscus schizopetalus
Family: *Malvaceae*, Mallow plants

Most important features: The hanging flowers have five petals that are light red, strongly divided, and curled. The midrib has many yellow anthers and five thin stigmatic branches at the end.
Growth structure: This bush grows up to 13 feet (4 m) high with thin, often drooping branches.
Leaves: The leaves are alternating and often grow in bunches at short lateral shoots. They are ovate to tapering, .8 to 4.8 inches (2 to 12 cm) long, and .4 to 2.8 inches (1 to 7 cm) wide with a serrated margin.
Flowers: The flowers grow individually. They hang on petioles that are up to 6.4 inches (16 cm) long. The flowers are 2 to 3.2 inches (5 to 8 cm) long with tiny epicalyx bracts at the bottom of the two to four dentate calyxes. The midrib is up to 3.6 inches (9 cm) long and often bent toward the side at the end.
Fruits: The fruits are tapering and up to 1.4 inches (3.5 cm) long. The capsules contain many seeds.
Occurrence: This is a very popular ornamental bush. Originally, it grew in tropical East Africa.
Other names: English "Fringed Hibiscus," "Japanese Lantern"; German "Fransen-Eibisch," "Koralleneibisch," "Zerschlitzter Roseneibisch."
Additional information: Because of its bizarre flowers, the coral hibiscus is used to breed new *Hibiscus* species almost as frequently as the Chinese hibiscus (see above). Many of the more than three hundred *Hibiscus* species (compare with photos above and below at right) are also used. The white color of the flower is preferred for interbreeding with the Hawaiian *H. arnottianus.* Although some of the inner bark fibers are used, as is also the case with the coral hibiscus, this is basically an ornamental plant. An exception is the Roselle Jamaica sorrel (*H. sabdariffa*). The calyxes are fleshy and purple during the flowering season. They are made into juice, jelly, or food colors.

Christ Thorn

Euphorbia milii
Family: *Euphorbiaceae*, Wolf's milk plants

Most important features: The Christ thorn has milky sap in all of its parts. The thick, squared branches have firm thorns arranged in pairs next to the leaves. The flowers have two red petals.
Growth structure: This bush grows up to 7 feet (2 m) high with firm, often crooked branches.
Leaves: The leaves are alternating, tapering to obovate, and 1.2 to 3.2 inches (3 to 8 cm) long.
Flowers: The inflorescences are up to 4 inches (10 cm) long and branched one to three times. Each has two, four, or eight flowers or partial inflorescences. Each has two spherical red bracts that grow up to .6 inch (1.5 cm) wide. Each also has five yellow shiny glands.
Fruits: The fruits are small and have three compartments. They are seldom seen.
Occurrence: This is a popular ornamental plant. It is also used as a hedge. It originated in Madagascar.
Other names: English "Crown of Thorns"; German "Christusdorn"; Spanish "Corona de Cristo."
Additional information: The name comes from the thorny crown of Christ. However, this is probably not the plant that was actually used because it was not known at that time in the Mediterranean region. The real Christ thorn was more likely *Paliurus spina-christi.* From these species have come several breeding specimens, including a dwarf one and one with yellow bracts. Despite the similarity, it is not the bracts but the five yellow glands that correspond to the organs of the scarlet plume (see page 162). The Christ thorn was originally at home in dry areas, as is true of the majority of the *Euphoria* species. In dry areas, many species show a growth structure similar to cacti, and they are often called cacti. Real cacti, however, have larger flowers with many petals, and they never have milky sap.

Guatemala Rhubarb

Jatropha podagrica
Family: *Euphorbiaceae*, Wolf's milk plants

Most important features: The trunk is thick and swollen at the bottom. The leaves have five lobes and accumulate at the top. The flowers are in erect false umbels. The last branching tends to be bright red like the petals.
Growth structure: The bush grows up to 5 feet (1.5 m) high, but it can reach 10 feet (3 m). It is slightly branched with fleshy, erect branches.
Leaves: The petioles are 4 to 12 inches (10 to 30 cm) long. The lobe is widest in the middle. The leaves have a wavy margin. The bottom side is often bluish.
Flowers: These are about .4 to .6 inch (1 to 1.5 cm) long. Most of them are male with six to ten yellow stamens. The few female flowers have three branched stigmas.
Fruits: The fruits are spherical capsules. They are .6 to .8 inch (1.5 to 2 cm) long and bluish green. When they ripen, they turn brown, and the three flaps open.
Occurrence: One finds the Guatemala rhubarb in gardens and in dry places. Originally, it was from Central America.
Other names: English "Gout Plant"; German "Rhabarber von Guatemala"; Spanish "Capa de Rey," "Ruibarbo."
Additional information: Although the Guatemala rhubarb can save water in its nodule, it still sheds its leaves during the dry period. It has milky sap in all of its parts. Many of the approximately 175 *Jatropha* species are cultivated. The coral plant (*J. multifida*) has similar flowers and deeply incised leaves. The seven to eleven slender fingers of the leaves can be lobed once more. The peregrina (*J. Integerrima*, see photo at right) has simple as well as lobed leaves on its thin branches. Its flowers are up to 1 inch (2.5 cm) long and often a deep pink. The Barbados nut (*J. curcas*) has small greenish flowers. The oil of its seeds is used to make candles and soap.

Desert Rose

Adenium obesum
Family: *Apocynaceae*, Dogbane

Most important features: The desert rose has a short, strikingly thick trunk. The flowers are pink to red with pink longitudinal stripes and a long, inner tube that is often yellow.
Growth structure: This plant usually grows up to 3 feet (1 m) tall but may reach 10 feet (3 m) high. The fleshy trunks have smooth, gray bark. All parts of the tree have a milky sap.
Leaves: The leaves accumulate at the ends of the branches. They are tapering and widest close to the tip. The leaves are 2.8 to 4.8 inches (7 to 12 cm) long, 1.2 to 3.2 inches (3 to 8 cm) wide, and slightly fleshy. On the upper side, they are dark green; on the bottom side, they are paler and dull. The tip is spherical or pointed.
Flowers: The flowers grow in groups on the leafless branches. They are up to 2 inches (5 cm) long. The corolla lobes are white with pink to crimson margins. They are twisted in the bud.
Fruits: The fruits are oblong to cigar-shaped. They are 9.6 inches (24 cm) long and .8 inch (2 cm) thick. Each pair grows at right angles to the petiole. The fruits open when ripe. The many seeds have a silky hair tuft on the ends.
Occurrence: The desert rose is popular as an ornamental plant in Asia and Africa. It originated in East Africa and spread from Ethiopia to South Africa.
Other names: English "Impala Lily," "Japanese Frangipani," "Mock Azalea," "Sabi Star"; German "Wüstenrose."
Additional information: While other ornamental plants need to be watered, the desert rose needs be protected from too much water. Because it originated in very dry areas, it stores the little water it needs in its thick trunk. If it has more water than it needs, it sheds its leaves. Although the plant is considered poisonous and its milky sap even serves as an arrow poison, it is eaten by many African wild animals for its water content.

Frangipani

Plumeria rubra
Family: *Apocynaceae*, Dogbane

Most important features: The frangipani has an abundance of milky sap in all its parts. The branches are thick, and the leaves accumulate at the ends. The corolla lobes are twisted like propellers.
Growth structure: This wide bush grows up to 20 feet (6 m) high and even higher in the wild.
Leaves: The leaves are tapering, stiff, dull, and leathery. They are 4.8 to 20 inches (12 to 50 cm) long and 1.2 to 6 inches (3 to 15 cm) wide.
Flowers: These grow in umbrella-shaped panicles at the branch tips. The corolla tube is .6 to 1 inch (1.5 to 2.5 cm) long, and the spread corolla lobes are 1 to 1.8 inch (2.5 to 4.5 cm) long. The flowers are red, pink, or white. On the bottom, they are often yellow.
Fruits: The fruits are tapering and slightly flattened. They are 4 to 12 inches (10 to 30 cm) long and .6 to 1.6 inch (1.5 to 4 cm) thick. Often, two grow together. Inside are many winged seeds.
Occurrence: The frangipani is frequently found in all tropical areas. Originally, it was from the Western Hemisphere.
Other names: English also "Temple Tree"; German "Frangipani," "Pagodenbaum"; French "Frangipanier"; Spanish also "Amapola," "Atapaima," "Flor de Cruz."
Additional information: The frangipani flowers almost all year round, except for a short leafless break. The wild specimen almost always seems to have white flowers with a yellow center. In gardens, the red specimen is found more frequently. Plain yellow species are only seen in cultivation. The Singapore graveyard flower (*P. obtusa*) always has white flowers with a yellow center, but it has shiny leaves with a rounded tip. Both species are symbols of eternal life, and so both are often planted in temple gardens and cemeteries. The flowers are used as offerings. On Hawaii, a flower behind the right ear signals the wearer is available; on the left, the wearer is already taken.

Scarlet Plume

Euphorbia fulgens
Family: *Euphorbiaceae*, Wolf's milk plants

Most important features: The scarlet plume has milky sap in all of its parts. The branches are thin and drooping. The leaves are slender and grow on long petioles. The flowers are orange red. The older ones have a green ovary standing out at one petiole.
Growth structure: This is a low bush that is not very woody and does not have many branches. The branches grow up to 7 feet (2 m) long.
Leaves: The leaves are alternating, tapering, and dark green. They are 2.4 to 5.6 inches (6 to 14 cm) long and .6 to 1 inch (1.5 to 2.5 cm) wide with steeply branching lateral veins and a joint between the petiole and leaf blade.
Flowers: The flowers grow in short panicles that point upward from a horizontal branch. The "flowers" are actually partial inflorescences, each with five orange red bracts that appear to be petals. They encompass one female and several male flowers.
Fruits: The fruits are small capsules with three segments. The fruits are seldom seen.
Occurrence: Although the scarlet plume is widely spread, it is not cultivated in gardens very often. Originally, it was from Mexico.
Other name: German "Mexikanische Wolfsmilch."
Additional information: The scarlet plume is a very attractive ornamental plant, which can often be found in flower shops, especially in winter. Besides the orange red wild specimen, a white specimen is also occasionally available. Among all of the approximately two thousand wolf's milk plants, scarlet plume is the best at imitating an individual flower using a partial inflorescence. Each of the five red bracts has a very strong shortened branch with several male flowers in its axil. These are short except for one single stamen. The central female flower has three red, stigmatic branches in its ovary.

Poinsettia

Euphorbia pulcherrima
Family: *Euphorbiaceae*, Wolf's milk plants

Most important features: The poinsettia has milky sap throughout all of its parts. The tips of the branches have many large, bright red leaves. The inflorescences are insignificant.
Growth structure: This plant grows up to 14 feet (4.5 m) high, often as a slightly branched bush.
Leaves: The leaves are alternating and often accumulate at the ends of branches. They are 2.8 to 12 inches (7 to 30 cm) long. The shape of the leaves depends on the specimen. Although most are tapering, the shapes are very diverse. Often the leaves have some indentations or lobes.
Flowers: The flowers grow between the leaves. They are small, yellow green, spherical formations. They have a one-sided yellow gland and tiny, red flower organs on the top. Even these are not the flowers, but partial inflorescences (see above).
Fruits: The fruits are spherical, green, and up to .6 inch (1.5 cm) long. Each fruit has three compartments.
Occurrence: The poinsettia is one of the most popular of all the tropical ornamental bushes. It originated in Mexico and Guatemala.
Other names: English "Christmas Flower," "Lobster Plant," "Mexican Flameleaf"; German "Weihnachtsstern"; Spanish "Bandera," "Guacamayo," "Flor de Noche Buena," "Flor de Pascua."
Additional information: The poinsettia needs nights that are longer than twelve hours in order to produce flowers. For this reason, it does well in northern areas around the Christmas season. As an ornamental plant, it is often kept small by using hormones. Specimens with pink, yellow, or white bracts are available on the market. The snows of Kilimanjaro (*E. leucocephala*) has much smaller leaves and is not suitable as an ornamental plant because of its strongly irritating milky sap. *E. cyathophora*, which is very similar to the poinsettia, is actually herbaceous and has red bracts only at the bottom. On the upper part, the bracts are green.

Annatto

Bixa orellana
Family: *Bixaceae*

Most important features: The petiole is slightly thickened at the stipula of the blade. The bottom of the leaf blade has three or five veins. The flowers are usually pink and have many stamens. The fruits are densely covered with white or red thorns.
Growth structure: As a bush, annatto grows up to 20 feet (6 m) tall. As a tree, it may reach 33 feet (10 m). The young branches are reddish brown and hairy.
Leaves: The leaves are alternating, light green, and oblong to heart-shaped. They hang on long petioles. The leaves are up to 10 inches (25 cm) long. The veins are often reddish.
Flowers: The flowers grow in panicles at the ends of the branches. They are about 2 inches (5 cm) long. Each has one hairy ovary.
Fruits: The fruits are ovate and slightly flattened. They are up to 1.6 inch (4 cm) wide with two flaps. Inside are up to fifty bright red seeds.
Occurrence: One finds annatto in gardens, but it often grows wild. Originally, it was from the Western Hemisphere.
Other names: English "Blood Tree," "Lipstick Tree"; German "Annatostrauch," "Orleansbaum"; French "Achiot," "Roucou"; Portuguese "Açafroa," "Urucú"; Spanish "Achiote," "Chaya," "Onoto."
Additional information: The fleshy seed coat of the annatto yields a bright red color that is used in lipsticks, soap, and foods such as cheese and margarine. The natives of the Western Hemisphere used it to color their hair and paint their skin.

Rose Cactus

Pereskia grandifolia
Family: *Cactaceae*, Cacti

Most important features: The rose cactus has light bunches of hair. Prickles grow in the axils of the leaves. The flowers have many pink or red petals and even more yellow stamens.
Growth structure: This is a firm bush that grows up to 16 feet (5 m) high. On rare occasions, it grows up to 23 feet (7 m) tall as a tree. The trunk is approximately 4 inches (10 cm) thick. Bunches of hair grow out of the thick branches. Each branch has from one to eight black thorns that are often .4 to 1.2 inch (1 to 3 cm) long.
Leaves: The leaves are alternating, elliptical, leathery, and pointed. They are 1.6 to 8 inches (4 to 20 cm) long and .8 to 2.8 inches (2 to 7 cm) wide.
Flowers: The flowers grow individually or at branched segments of a shoot. The flowers are 1.6 to 2.8 inches (4 to 7 cm) long and have small deciduous leaves. The flowers have numerous whorls of bracts. The outer ones are cone-shaped and green. However, most of them are multicolored. The pistil has several white stigmas.
Fruits: These pear-shaped, yellow berries are 2 to 2.4 inches (5 to 6 cm) long. Some thickened leaves of the perianth remain on the surface.
Occurrence: The rose cactus is widely spread as an ornamental bush. It originated in Brazil.
Other names: English "Wax Rose"; German "Rosenkaktus"; Spanish "Ñajú de Espinas."
Additional information: Despite its leaves, the rose cactus is a real cactus, which is obvious when one examines its thorns and flowers. Foliaged cacti are a tiny minority among the approximately 1,400 cacti species. For example, *Pereskia* has sixteen species, and *Maihuenia* has two. The climbing Barbados gooseberry (*P. aculeata*) is planted more frequently. It has slightly smaller flowers and two firm, curled horns at the base of the leaves. *P. corrugata* has orange flowers that look like roses before they open. It is a popular ornamental bush.

Sleeping Hibiscus

Malvaviscus arboreus
Family: *Malvaceae*, Mallow plants

Most important features: The flowers are bright red but may also be pink. They are 1.2 to 2.4 inches (3 to 6 cm) long. The five petals are twisted together into a tube. The midcolumn with many stamens and the pistil with ten stigmatic branches emerge from the tube.
Growth structure: The sleeping hibiscus grows up to 13 feet (4 m) high as an erect bush. It may also grow up to 33 feet (10 m) high as a tree.
Leaves: The leaves are alternating, very variable, and ovate to tapering. They are 2.8 to 8.4 inches (7 to 21 cm) long and 1.6 to 6.4 inches (4 to 16 cm) wide. Often slightly lobed, they are spherical to heart-shaped at the bottom and tapering with serrated margins. They are hairy on both sides. The thin leaves are herbaceous with emerging veins.
Flowers: The individual flowers may be erect or drooping. They are slender with hairy, epicalyx bracts at the bottom of the calyx, which is .4 to .8 inch (1 to 2 cm) long and has two to five lobes.
Fruits: The fruits are flattened and spherical. They are 3.2 to 6 inches (8 to 15 cm) long and reddish to bluish. The fruits are slightly fleshy and fall apart, creating five partial fruits.
Occurrence: Frequently planted as an ornamental bush, the sleeping hibiscus originally spread from the southern United States to South America.
Other names: English "Turk's Cap," "Wax Mallow"; German "Beerenmalve," "Wachsmalve," "Schlafeibisch"; Spanish "Amapola," "Manzanita," "Monacillo," "Quesillo," "Papito de Monte," "Tulipancillo."
Additional information: The flowers of the sleeping hibiscus never appear to be completely open. Actually, that is correct. They never do open completely because the tube has an important function. When hummingbirds visit, the way they insert their bill to drink the nectar (see photo at left) causes it to touch first the stigmas, then the stamens. When it flies to the next flower, it transfers the pollen.

Trailing Abutilon

Abutilon megapotamicum
Family: *Mavaceae*, Mallow plants

Most important features: The flowers droop. They have a red calyx and winged-shaped edges. The petals are light yellow. The one midcolumn grows out of many stamens that are dark violet to black.
Growth structure: This plant grows up to 7 feet (2 m) high with thin, drooping branches. Sometimes, the branches climb onto other bushes.
Leaves: The leaves are alternating with a triangular tip and a serrated margin. They are 2 to 4 inches (5 to 10 cm) long and .8 to 1.4 inch (2 to 3.5 cm) wide with a heart-shaped base. The stipules are .4 inch (1 cm) long and .2 inch (5 mm) wide.
Flowers: The flowers grow individually and hang on petioles that are 1 to 2.4 inches (2.5 to 6 cm) long. The calyx is 1.2 inch (3 cm) long and heart-shaped in the bud. The petals are about .6 inch (1.5 cm) long and tower above the calyx. The stamen column emerges about the same length from the corolla.
Fruits: The fruits have five parts that open individually toward the top, each with two to six seeds.
Occurrence: Frequently used as an ornamental plant in gardens, the trailing abutilon sometimes grows wild. The plant originated in the southern part of Brazil.
Other name: German "Dreifarbige Schönmalve."
Additional information: The trailing abutilon is one out of a hundred *Abutilon* species spread over all warm areas. Many are ornamentals. Some, however, like *A. indicum*, are weeds and contain a strong insect poison. *A. theophrasti*, which produces the fibers for China jute, grows wild. Among the ornamental plants, one can often find specimens with spotted leaves. With *A. pictum*, which has a green calyx and an orange red, bell-shaped corolla, the spotted specimen (*thompsonii*) is seen most frequently. The spots are caused by a virus that does not affect the plant in any other way.

Firebush

Streptosolen jamesonii
Family: *Solanaceae*, Nightshade

Most important features: The flowers come in different shades of orange and grow in inflorescences. They have five corolla lobes. The lower one is wider; the two upper ones are more slender than the two on the side, and often they are curled under slightly.
Growth structure: This bush grows up to 8 feet (2.5 m) high with long, thin, hairy branches that droop. Sometimes the branches are trained as espaliers.
Leaves: The leaves are alternating, ovate to elliptical, and .8 to 2 inches (2 to 5 cm) long. They are hairy on the bottom side. Between the veins, they are slightly vaulted.
Flowers: The flowers grow in dense panicles that are up to 8 inches (20 cm) long. They are often yellow when opening and almost red when withering. The calyx is about .4 inch (1 cm) long and has five unequal indentations. The crooked corolla tube has delicate hairs and is 1.2 to 1.6 inch (3 to 4 cm) long. These corolla lobes are spherical and slightly folded lengthwise. The four stamens are located at the top of the tube.
Fruits: The fruits have two indistinct parts with many seeds. For a long time, the fruits are encompassed in the calyx.
Occurrence: One finds the firebush in private houses and in public gardens. Originally, it was from the western part of South America.
Other names: English "Marmalade Bush," "Orange Browallia," "Yellow Heliotrope"; German "Marmeladenstrauch."
Additional information: The firebush blooms almost all year round. Because it originated at high altitudes, it can easily withstand cool nights. It can even be put outside in a pot in summer. Because it grows very rapidly and flowers its first year, it is sometimes kept as an annual. In earlier times, it was considered to be part of the *Browallia* genus. The two species of *Browallia* are herbaceous and have similarly shaped flowers, but the flowers are violet or even white.

Carnival Bush

Ochna serrulata
Family: *Ochnaceae*

Most important features: The leaves are small with delicate indentations on the margins. The flowers are golden yellow with many stamens. The fruits are flower-shaped with stony, black fruits on a bright red flower bed.
Growth structure: The carnival bush is richly branched and grows up to 10 feet (3 m) high. As a small tree, it may reach 20 feet (6 m). The branches are strewn with tiny pores.
Leaves: The shiny leaves are alternating and elliptical. They are 2 inches (5 cm) long and just .4 inch (1 cm) wide.
Flowers: These grow individually or occasionally in pairs on short stipules. They are up to .8 inch (2 cm) long with five green sepals, five petals, and an ovary with five to seven parts and one common pistil.
Fruits: The fleshy flower bed is swollen during the flowering season. At the bottom, it is enlarged and up to .8 inch (2 cm) long with curled sepals on the margins and the thickened remainders of stamens. The fruits are spherical and have stone pits that are up to .4 inch (1 cm) long.
Occurrence: Frequently found in gardens, the carnival bush originated in the eastern part of South Africa.
Other names: English "Mickey Mouse Tree"; German "Mickymausbusch," "Ochna."
Additional information: The carnival bush is often cultivated for its bizarre fruits. Some people find the fruits reminiscent of Mickey Mouse with their spherical, black partial fruits. The remaining eighty-five *Ochna* species are often relatively small bushes. *O. arborea* grows much larger, up to 40 feet (12 m) high. Its heavy wood is used for tool handles and fence posts. People also use the wood to whittle into amulets. A powder of its bark is inhaled for headaches, and the roots have played a role in tribal medicines.

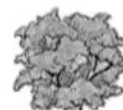

Malayan Dillenia

Dillenia suffruticosa
Family: *Dilleniaceae*, Rose Apple

Most important features: The large, elliptical leaves have firm lateral veins, running almost straight into the dentate margins. The flowers are bright yellow and very large. The fruits are similar to flowers, but they are red.
Growth structure: This wide bush grows up to 33 feet (10 m) high.
Leaves: The leaves are alternating. They are approximately 10 inches (25 cm) long and 4.8 inches (12 cm) wide with up to twenty pairs of veins. The leaves are bent just before the margin.
Flowers: The flowers are up to 6 inches (15 cm) wide with five lightly reddish sepals. The flowers have five large, delicate, and slightly wavy stamens and five to eight white ray flowers. The flowers grow in short simple racemes. They are each only open for one day.
Fruits: The fruits are star-shaped and up to 2.4 inches (6 cm) in diameter with curved sepals reflexed. The five to eight points of the star are created by bright red carpels. The star in the center is a white funiculus. Inside are red black seeds.
Occurrence: The Malayan dillenia is a very common plant in Malaysia and Indonesia. It often grows in wild bushland in very humid soil. It grows wild in the Caribbean.
Other names: Malayan "Simpoh Air"; German "Simpurstrauch."
Additional information: The sixty species of the *Dillenia* genus are spread from Madagascar to Fiji and from the Himalayas to Australia. Most of the species can be found in the Indo-Malaysian region. Despite their decorative flowers, which bloom almost all year long, only a few species are cultivated (see page 106). In many places, they are considered weeds because they rapidly cover fields. *Dillenia excelsa* is easily recognizable because of its yellow flowers and violet stamens. Each branch that hits the ground is supposed to take root again.

Coast Hibiscus

Hibiscus tiliaceus
Family: *Malvaceae*, Mallow plants

Most important features: The leaves are almost perfectly round. The large flowers are yellow to orange. In the middle, they are very dark purple. The calyx is enclosed in a large epicalyx.
Growth structure: This is a wide bush that is usually not shaped like a tree. It grows up to 33 feet (10 m) high. It has crooked branches that may rest on the ground. Young branches have soft hair.
Leaves: The leaves are alternating and pointed. They grow up to 8 inches (20 cm) long with a heart-shaped base and five to nine main veins. On the upper side, the leaves are a dull green; on the bottom side, they are paler with soft hair. The stipules grow up to 1.2 inch (3 cm) long.
Flowers: These are up to 4.8 inches (12 cm) long with eight to eleven lobed epicalyxes. The midcolumn grows up to 2 inches (5 cm) long with numerous filaments branching off of it. The pistil is about .4 inch (1 cm) longer than the midcolumn, and the stigmas are dark.
Fruits: The fruits are ovate capsules up to 1.4 inch (3.5 cm) long. They are pointed, yellow, and hairy with five flaps. They seeds are approximately .2 inch (5 mm) long.
Occurrence: One finds the coast hibiscus in the tidal areas of tropical coasts.
Other names: English "Hau Tree," "Mahoe Tree," "Wild Cotton Tree"; German "Lindenblättriger Eibisch"; Spanish "Majagua."
Additional information: The coast hibiscus has spread almost around the world because its seeds float. The bark contains fibers from which robes, nets, and skirts are produced. The crooked branches are used as boat covers. The flowers live for only one day. They open in the morning a bright yellow but become more and more orange as the day progresses. In the evening, they wilt and turn dark red brown. One can differentiate between the coast hibiscus and the very similar cork tree (see page 102) by the separated dark spots within the flowers, the cup-shaped calyx, and the light stigmas of the cork tree.

Yellow Bauhinia

Bauhinia tomentosa
Family: *Caesalpiniaceae*, Carob plants

Most important features: The leaves are divided into two spherical lobes. The flowers are yellow; the upper or inner of the five petals has a dark purple (almost black) spot on the bottom.
Growth structure: This is a dense bush with drooping branches. As a tree, it rarely reaches 23 feet (7 m).
Leaves: The leaves are alternating and 1 to 3.2 inches (2.5 to 8 cm) long. At the bottom, they are spherical to heart-shaped. Each lobe has three veins.
Flowers: These grow individually or in small flowering panicles at the ends of branches. They are 1.2 to 2.8 inches (3 to 7 cm) long and bell-shaped with five stamens inside.
Fruits: These grow up to 4.4 inches (11 cm) long and .8 inch (2 cm) wide. They are flattened, pale brown, and velvety or feltlike. When ripe, they break into two parts.
Occurrence: Often used as an ornamental bush, the yellow bauhinia is widely spread in tropical and subtropical Africa and Asia.
Other names: English "Bush Neat's Foot"; German "Gelbe Bauhinie."
Additional information: The yellow bauhinia is one of about three hundred *Bauhinia* species, most of which have very striking flowers. Many are cultivated as ornamental plants. Almost all of them exhibit the typical leaf, which is divided into two parts, almost to the petiole. This is true of bushes, trees (page 98), or vines (page 202). Many species can grow as trees or bushes, and in the case of the yellow bauhinia, as a bush or as a vine. Most of these plants have reddish or white flowers with spreading petals. Yellow flowers with overlapping petals (see photo at right) are rare. With the equally bushy *B. cumingiana* and some climbing species, the young flowers are yellow at first, but change color during the flowering season, becoming more and more orange.

Be-Still Tree

Thevetia peruviana
Family: *Apocynaceae*, Dogbane

Most important features: The plant has milky sap in all parts. The leaves are very narrow. The flowers are yellow (occasionally reddish) and funnel-shaped.
Growth structure: The be-still tree is richly branched and grows up to 30 feet (9 m) high. When cultivated, it barely reaches 13 feet (4 m).
Leaves: The leaves are alternating and tapered at both ends. They are 2.8 to 6 inches (7 to 15 cm) long and .2 to .6 inch (0.5 to 1.5 cm) wide. The leaves are light green and leathery.
Flowers: The flowers grow individually or a few together. Each flower has five petals. The flowers are 1.2 to 2.4 inches (3 to 6 cm) long. The tube is narrow, then suddenly widens and twists the spherical corolla lobes. The corolla lobes are up to 1.6 inch (4 cm) long.
Fruits: The fruits are flattened, elliptical to spherical, and pyramid-shaped. They are 1 to 1.2 inch (2.5 to 3 cm) long and 1.2 to 1.6 inch (3 to 4 cm) wide. When ripe, they are dark red with two to four seeds.
Occurrence: Found in all tropical areas. It originated in the Western Hemisphere.
Other names: English "Lucky Nut," "Trumpet Flower," "Yellow Oleander"; German "Schellenbaum," "Tropischer Oleander"; French "Arbre à Lait," "Noix-Sirpent," "Oléandre Jaune"; Portuguese "Loandro Amarelo"; Spanish "Adelfa Amarilla," Cabalonga," "Campanilla," "Cascabel," "Chilco," "Retama."
Additional information: The be-still tree is only remotely related to the real oleander (*Nerium oleander*) of the Mediterranean area; it is, however, similarly poisonous. It contains a deadly heart glycoside so strong that eating one fruit can be fatal. Strongly diluted, the milky sap serves as a treatment for fever, toothache, and stomach ulcers in Mexico. Parts of the brittle pits of the fruit are sometimes worn as jewelry or rattles. In Asia, the flowers are often part of Buddhist and Hindu flower offerings.

Yellow Angel's Trumpet

Brugmansia aurea
Family: *Solanaceae*, Solanum

Most important features: The flowers are very large and trumpet-shaped. They droop with five delicate, curled, pointed ends.
Growth structure: The yellow angel's trumpet often grows very wide and bushy. It has thick, soft branches and usually grows up to 13 feet (4 m) high. On rare occasions, it reaches a height of 33 feet (10 m).
Leaves: The tapering leaves, 6 to 10 inches (15 to 25 cm) long, are ovate and often slightly wavy. Young plants have dentate margins. Small leaves frequently grow in the axils of large ones.
Flowers: The flowers grow individually or two together. They have a long, tubular calyx. The corolla tube is yellow to almost white and 6 to 12 inches (15 to 30 cm) long. Hidden in the tube are a long pistil and five stamens.
Fruits: The fruits are tapering, ovate, smooth, and woody. They contain many wedge-shaped seeds.
Occurrence: The yellow angel's trumpet is a popular garden plant. Originally, it was from the northern Andes.
Other names: German "Totentrompete," "Gelbe Engelstrompete"; Spanish "Borrachero."
Additional information: The yellow angel's trumpet owes its name to its strong poison. Extracts of the leaves and seeds have served the shamans of South America as hallucinogenic drugs, used to get in touch with the gods. The Spanish name, *borracho*, means drunk and refers to this particular use. In addition to the yellow angel's trumpet, there are other *Brugmansia* species in cultivation. They are similarly used, especially *B. arborea*, *B. suaveolens*, and *B. versicolor*, which have white flowers. The crossbreeds *B.x candida* and *B. x insignis*, which are called angel's tears and *B. sanguinea*, the red angel's trumpet, have a yellow corolla tube and an orange red and yellow striped corolla lobe. All were originally placed in the *Datura* genus; however, that genus is herbaceous and produces thorny fruits.

Thai Caper

Capparis micracantha
Family: *Capparidaceae*, Caper plants

Most important features: The flowers grow in groups of up to six buds in one row on the upper side of the petioles. They have long filaments and white petals. The upper one has a yellow to purple spot. The ovary grows on a petiole.
Growth structure: The bush grows up to 20 feet (6 m) high. It may also grow as a small tree or as a climbing plant. The branches droop and have short thorns.
Leaves: The leaves are alternating and elliptical to tapering. They are 3.2 to 7.2 inches (8 to 18 cm) long and 1.6 to 3.2 inches (4 to 8 cm) wide. The leaves are shiny and light green.
Flowers: The flowers have four petals that are .4 to 1 inch (1 to 2.5 cm) long and about .2 inch (5 mm) wide. The stamens and petiole of the ovary are about 1.2 inch (3 cm) long. The ovary is green.
Fruits: The leathery fruits are spherical to tapering. They are 2.6 inches (6.5 cm) long and 1.8 inches (4.5 cm) thick. When ripe, they are yellow to red with a juicy flesh and many seeds.
Occurrence: Originally from Southeast Asia, the Thai caper is often found there in gardens in semishade.
Other name: German "Thailändischer Kapernstrauch."
Additional information: Like most of the approximately 250 *Capparis* species, the Thai caper is pollinated by nighthawks. These fly from flower to flower, touching both stamens and stigma. The long petiole of the ovary is a family characteristic. The Thai caper is related to the real caper bush (*C. spinosa*), found in the Mediterranean region; however, the Thai caper is not suitable as a spice. In order to develop the typical aroma, the buds of the real caper must wilt shortly after they are harvested. Then, they are salted and marinated. Like most other *Capparis* species, the real caper bush has white petals with no dark spots.

Tea Plant

Camellia sinensis
Family: *Theaceae*, Tea plants

Most important features: The leaves have dentate margins. The flowers have five to seven sepals and petals, many stamens, and a distinct pistil that is split into three parts.
Growth structure: The tea plant grows up to 16 feet (5 m) high as a bush. It may grow up to 65 feet (20 m) high as a tree. When cultivated, it is usually kept at about 3 feet (1 m) tall.
Leaves: The leaves are alternating and ovate to tapering. They are up to 4.8 inches (12 cm) long. The leaves of the two main specimens are slightly different (see below).
Flowers: The flowers grow individually or in groups of up to four together. They are 1 to 1.6 inches (2.5 to 4 cm) long. The sepals are wide and green. They remain on the fruit. The petals are white or yellowish to slightly pink. The stamens are small and spherical.
Fruits: The fruits are ovate, leathery, pointed, and up to 1.2 inch (3 cm) long. They have three capsules, each with one thick, spherical, dark brown seed.
Occurrence: The tea plant has spread widely in cultivation. Typically it is planted in the highlands as dense hedges. It originated on the southern flank of the Himalayan range.
Other names: German "Teestrauch"; French "Arbre à Thé"; Portuguese "Chá da Índia"; Spanish "Arbol de Té."
Additional information: More than 2.5 million tons of tea reach the market every year. That makes tea an important part of the worldwide economy. The Chinese have cultivated it for more than five thousand years. Chinese tea (var. *sinensis*) is moderately frost-resistant and has slightly smaller, firm, leathery, strongly dentate leaves. In the nineteenth century, growers discovered that Assam tea (var. *asamica*) is sensitive to frost and has slightly larger, thinner, and indistinctly dentate leaves. Black tea is created by wilting and fermenting the leaves after the harvest. With green tea, the decomposition is stopped early by heating and drying, so that many of the original substances remain.

Beach Naupaka

Scaevola tacccada
Family: *Goodeniaceae*

Most important features: The leaves are rosette-like and accumulate at the ends of branches. The flowers have a split tube. At the top, the five corolla lobes point sideways and toward the ground.
Growth structure: The beach naupaka is a dense, wide bush that grows up to 10 feet (3 m) high with firm, arch-shaped rising branches. It may also grow as a tree that is up to 23 feet (7 m) high.
Leaves: The leaves are light green, shiny, and slightly fleshy. They are 4.8 to 10.4 inches (12 to 26 cm) long and 2 to 4 inches (5 to 10 cm) wide. They are spatula-shaped just before the spherical tip. Toward the bottom, they are wedge-shaped.
Flowers: The flowers grow in small groups between the leaves. They are .8 to 1 inch (2 to 2.5 cm) long. The flowers are pale yellow to white with rapidly fading violet lines. The pistil is firm with a wide stylar head that is bent toward the ground.
Fruits: The fruits are white and up to .6 inch (15 mm) long. The pits are stony. The fruits are crowned by the remainders of the perianth and often appear to be split into two halves or grooved.
Occurrence: One finds the beach naupaka on the beaches of the Indian and Pacific Oceans. As hedge plants, they are planted in sandy ground.
Other names: English "Half-Flower," "Sea Lettuce"; German "Fächerstrauch."
Additional information: Like other beach plants, the beach napauka has spread without human help. Birds eat its pits and spread them. Additionally, the pits have a corklike outer layer and can float in sea water for long periods without losing their ability to germinate. The closely related false napauka (*S. aemula*) has violet, light blue, or white flowers with a yellow center.

Croton

Codiaeum variegatum var. pictum
Family: *Euphorbiaceae*, Wolf's milk plants

Most important features: The leaves are often red, yellow, and green with spots or speckles. They have obvious veins. The leaves are leathery with smooth edges and may be simply lobed or may have no lobes. The flowers are relatively insignificant and unisexual.
Growth structure: This is an erect bush that grows up to 20 feet (6 m) high. When cultivated, it rarely reaches more than 10 feet (3 m).
Leaves: The leaves are alternating and widely variable in shape. They also vary in size from 2 to 16 inches (5 to 40 cm) long. The leaves are elliptical to ribbon-shaped and occasionally wavy.
Flowers: The flowers grow in drooping simple racemes up to 10 inches (25 cm) long. Each flower has five tiny sepals. Male flowers have long petioles with five or six tiny, dentate petals. They also have many glands and fifteen to thirty-five stamens. Female flowers are sessile. The ovaries have three long, curled stigmas.
Fruits: The fruits are spherical and pale green with three indistinct lobes. They grow up to .4 inch (1 cm) long. When ripe, they fall, breaking into three segments. Each has one brown, spotted seed.
Occurrence: Croton is one of the most popular ornamental bushes. It is also sold as a houseplant. Although the precise place of origin is uncertain, the most likely site is the Indo-Malaysian area.
Other name: German "Wunderstrauch."
Additional information: The leaves of the croton are similar to those of the beefsteak plant (see photo below); however, croton leaves have a dentate margin, are thinner, and are herbaceous. Depending on the shape and color of the leaves, one can differentiate among numerous cultural forms. In New Guinea, they sometimes serve as the frontier of tribal areas. In Southeast Asia, people often eat the young branches and leaves as vegetables. The oil of the seeds may have a strong laxative effect.

Beefsteak Plant

Acalypha wilkesiana
Family: *Euphorbiaceae*, Wolf's milk plants

Most important features: The leaves are wide and ovate to heart-shaped. They have a serrated margin. The leaves have at least two colors. Red is usually the dominant color, but the leaves may have a light margin.
Growth structure: This is a wide, dense bush that grows up to 16 feet (5 m) tall.
Leaves: The tapering leaves are alternating and herbaceous with clearly emerging veins on the bottom side. They are 1.6 inch to 8.8 inches (4 to 22 cm) long and .8 to 6 inches (2 to 15 cm) wide.
Flowers: The flowers are insignificant. The male ones grow in long, thin, drooping spikes that grow up to 8 inches (20 cm) long, The female ones grow on short, loose, erect spikes.
Fruits: The fruits are tiny and spherical. When they are ripe, they fall, breaking into three parts.
Occurrence: One finds this bush growing in gardens and public areas. Originally, it was probably from Fiji.
Other names: English "Copperleaf," "Jacob's Coat," "Fire Dragon," "Match-Me-If-You-Can"; German "Schillerndes Nesselblatt"; Spanish "Capa Roja."
Additional information: The beefsteak plant is popular because of its colored leaves; the flowers and fruits are insignificant. Innumerable color varieties are available. The most popular specimens are macafeeana (light crimson leaves with dark spots), macrophylla (wine red to purple leaves), marginata (purple to olive-brown leaves and pink or orange margins), and mosaica (spotted leaves that may have three to four colors, including olive brown, purple, crimson, copper red, orange, and green). At one time, the specimen godseffiana was considered an individual species because its leaves are not red, but rather green with light yellow or white margins. In addition, specimens of many species have wavy leaves.

Butterfly Pea

Clitoria ternatea
Family: *Fabaceae*, Papilionaceous plants

Most important features: The leaves are pinnately divided. The flowers have one large petal, which points toward the ground, creating an oval shape. The upper third of this petal has two pairs of much smaller petals.
Growth structure: The plant is herbaceous with thin shoots that grow up to 20 feet (6 m) high.
Leaves: The leaves are alternating. They are 2.4 to 6.8 inches (6 to 17 cm) long. The five to seven elliptical pinnules are .8 to 2.8 inches (2 to 7 cm) long and .4 to 1.6 inch (1 to 4 cm) wide.
Flowers: The flowers usually grow individually at the bottom with two ovate little leaves that are .2 to .4 inch (0.5 to 1 cm) long. The calyx is .6 to 1 inch (1.5 to 2.5 cm) long; half of it is stunted. The large petal is 1.4 to 2 inches (3.5 to 5 cm) long and 1 to 1.6 inch (2.5 to 4 cm) wide. The flowers are blue to violet or rarely white. They have a yellow spot in the middle of the white margin. The small petals are slightly wavy. The two outer ones are only half as long as the flower. The inner ones are only half the size of the outer ones.
Fruits: The fruits are green and bean-shaped. They are 2 to 4.8 inches (5 to 12 cm) long, about .4 inch (1 cm) wide, and slightly hairy. Each fruit has six to ten seeds.
Occurrence: One can find this plant growing anywhere in the tropics, but especially along sidewalks. The place of origin is uncertain, but the most likely area is the Western Hemisphere.
Other names: English "Blue Pea," "Cordofan Pea"; German "Schmetterlingserbse"; Spanish "Azulejo," "Conchitas," "Papito," "Zapatico de la Reina."
Additional information: The flowers of the butterfly pea are twisted 180 degrees in comparison with those of other plants in the same family, such as the pea. They react like litmus paper, turning red in the presence of acids and blue in the presence of alkalines. In Asia, the flowers are used as a dye for coloring rice or pastry. The name, *Clitoria*, refers to the flower. With a lot of imagination, the flower is supposed to be reminiscent of parts of the female genitalia.

Cathedral Bells

Cobaea scandens
Family: *Polemoniaceae*, Greek valerian

Most important features: The leaves are alternating and pinnately divided. Often a tendril replaces the terminal pinna. The flowers are very large, dark violet, and bell-shaped.
Growth structure: This is a large, herbaceous vine, which grows 10 to 26 feet (3 to 8 m) in one year. It may grow to a height of 65 feet (20 m).
Leaves: The leaves have three (occasionally only two) pinnules. The lower one is sessile and has a dull, heart-shaped bottom. The upper ones have petioles that are up to .8 inch (2 cm) long. They are elliptical to obovate and tapering. These leaves are 1.4 to 5.2 inches (3.5 to 13 cm) long and .6 to 2.4 inches (1.5 to 6 cm) wide. The tendrils are often branched at the tip and have a hook.
Flowers: These grow individually. The calyx is 1 to 1.4 inch (2.5 to 3.5 cm) long. Half of it is stunted and has five wing-shaped emerging edges. The corolla is 1.8 to 2.6 inches (4.5 to 6.5 cm) long with wide, spherical lobes. Before unfolding, the leaves are greenish white, but they seldom remain that color. The stamens and pistil are located at the bottom of the corolla. They are slightly bent and point upward.
Fruits: The fruits are elliptical to tapering. They are 2 to 3.6 inches (5 to 9 cm) long and sit on a lobed disk. When they are ripe, they open with three flaps, releasing large, winged seeds.
Occurrence: Originally from the Western Hemisphere, it is widely cultivated and sometimes spreads beyond the desired area.
Other names: English "Cup-and-Saucer Vine," "Mexican Ivy"; German "Glockenrebe," "Krallenrebe."
Additional information: Cathedral bells produces a lot of nectar, which attracts bats to pollinate the flowers. Because it grows naturally at about 6,500 feet (2,000 m), it can withstand cool nights, but not frost. It can be grown as a summer annual, but to flower, it must be presprouted at about 68°F (20°C) by the end of February and planted outdoors only after the danger of frost has passed.

Pink Tecoma

Podranea ricasoliana
Family: *Bignoniaceae*, Trumpet tree plants

Most important features: The leaves are set opposite, unpaired, and pinnately divided. The pinnules have a serrated margin. The flowers are pink with darker longitudinal stripes in the throat.
Growth structure: This plant grows up to 16 feet (5 m) high. It climbs with relatively thin, but woody branches.
Leaves: Leaves up to 10 inches (25 cm) long have five to eleven ovate pinnules 1 to 1.6 inch (2.5 to 4 cm) long and .6 to .8 inch (1.5 to 2 cm) wide.
Flowers: Flowers grow in panicles at the ends of branches. The pale calyx, shaped like a wide bell, is .6 to .8 inch (1.5 to 2 cm) long. About half of it is divided into five tapering lobes. The corolla is 2.4 to 3.2 inches (6 to 8 cm) long with five lobes. The corolla tube is pale pink to yellowish white. Inside, it has pink stripes and spots. At the bottom, it is narrow, then bell-shaped with two long stamens and two short ones.
Fruits: The fruits area almost cylindrical and 10 to 15 inches (25 to 35 cm) long. They are leathery. When ripe, the two flaps open, spreading many winged seeds.
Occurrence: Frequently, one sees these plants growing on house walls. Originally, it grew on Port St. John in South Africa.
Other names: English "Pink Trumpetvine," "Port St. Johns Creeper"; German "Rosa Trompetenwein."
Additional information: The pink tecoma loses its leaves in cold weather. However, it survives frost up to 23°F (–5°C). The similar species, *P.brycei* (Zimbabwe creeper), is slightly more sensitive. The pinnules are more slender, the calyx has longer cusps, and the corolla tube is broader at the bottom. At one time, both species were placed in the *Pandorea* genus. The most frequent species of the *Pandorea* genus, *P. jasminoides*, is similar, but one can easily differentiate it by its pinnules, which have continuous margins, and by its dark pink throat.

Flame Vine

Pyrostegia venusta
Family: *Bignoniaceae*, Trumpet tree plants

Most important features: The leaves are opposite. They grow in groups of three or with a three-part tendril instead of the terminal pinna. The flowers are orange with a long tube and five cusps. The two upper ones are connected halfway.
Growth structure: The plant grows up to 50 feet (15 m) high, but may creep along the ground when cultivated. The branches are slightly squared.
Leaves: The leaves grow two or three pinnules on a petiole. The pinnules are 1.6 to 4 inches (4 to 10 cm) long and .8 to 2 inches (2 to 5 cm) wide. They are ovate to tapering and shiny.
Flowers: Flowers grow in panicles at the ends of the branches and are 2 to 3 inches (5 to 7.5 cm) long. The corolla cusps curl under at the ends. The four stamens and the pistil tower above the tube.
Fruits: The fruits are oblong and flattened with two flaps. They are 10 to 12 inches (25 to 30 cm) long and .4 to .6 inch (1 to 1.5 cm) thick.
Occurrence: In many areas, the flame vine is grown as an ornamental plant. It originated in Brazil and Paraguay.
Other names: English "Golden Shower," "Orange Creeper," "Sweetheart Vine"; German "Feuerranke"; French "Liane Aurore"; Portuguese "Cipó de São João"; Spanish "Chiltote," "Chorro de Oro," "San Carlos," "Triquitraque."
Additional information: The flame vine prefers locations in the bright sun and is quite sensitive to cold. Other climbing plants of the same family can endure light frost. These plants have orange to red flowers. The cape honeysuckle (*Tecoma capensis*) from South Africa has even smaller flowers. The trumpet creepers (*Campsis grandiflora* from China, *C. radicans* from North America, and their crossbreed *C. x tagliabuana*) have much larger flowers that are 2 to 3.6 inches (5 to 9 cm) wide. All of these species have pinnately divided leaves, five to nine of which have serrated edges but no tendrils.

Blue Bird Vine

Petrea volubilis
Family: *Verbenaceae*, Vervain plants

Most important features: The leaves are set opposite and are very rough, like sandpaper. The flowers have five long, light blue to violet sepals and a smaller, darker corolla with unequal lobes.
Growth structure: This is a winding vine that grows up to 43 feet (13 m) long. Sometimes, it grows as a bush with drooping branches.
Leaves: The leaves are elliptical, light green, very rough, and often slightly wavy with emerging veins. They grow 2.4 to 8.4 inches (6 to 21 cm) long and .8 to 4.4 inches (2 to 11 cm) wide.
Flowers: The flowers grow in drooping simple racemes that are 3.2 to 12 inches (8 to 30 cm) long. The sepals grow up to .8 inch (2 cm) long. At the bottom, the sepals are connected to a short tube that continues beyond the cusps and is oblong before blooming. It remains after the flower opens. The corolla is up to .6 inch (15 mm) wide with a tube that is .3 inch (8 mm) long. The flowers do not last long.
Fruits: This plant has small stone fruits with little flesh and one or two pits. They fruits are entirely enclosed in the calyx.
Occurrence: One finds the blue bird vine in gardens. It is native to Central America and the Antilles.
Other names: English "Purple Wreath," "Queen's Wreath," "Sandpaper Vine"; German "Purpurkranz"; Spanish among others "Adolfina," "Corona de la Reina," "Estrella Azul," "Flor de Jesús," "Penitente," "Santa María," "Soltero."
Additional information: The blue bird vine is among the most popular ornamental plants because of its unusual color and the durability of its sepals. Many of its Spanish folk names refer to saints or to the Church, perhaps because its violet color is associated with the Church. One also sees white flowers, but these are rare. Others of the approximately thirty *Petrea* species always have white calyxes, such as the tree-shaped *P. glandulosa*.

Bengal Clockvine

Thunbergia grandiflora
Family: *Acanthaceae*, Acanthus plants

Most important features: The leaves are set opposite. The leaf margins are dentate to weakly lobed. The flowers are large and light blue to light violet with a yellow throat. There are four stamens. The anthers are about .2 inch (5 mm) long with spurs directed backward.
Growth structure: This is a winding plant with shoots that becomes woody. It grows up to 100 feet (30 m) long.
Leaves: The pointed leaves, heart-shaped to ovate, are 2.8 to 8 inches (7 to 20 cm) long and 1.2 to 7.2 inches (3 to 18 cm) wide with five to seven veins running from the heart-shaped bottom.
Flowers: Flowers grow in the axils of small leaves or in drooping simple racemes that are up to 7 feet (2 m) long. The margin of the calyx is almost continuous. It has dark spots and is covered by two stunted, pale green bracts. The calyx is .8 to 1.6 inch (2 to 4 cm) long. The corolla is 2 to 4 inches (5 to 10 cm) wide with five lobes. The upper two lobes slightly overlap; the one at the bottom is vaulted in the middle. The corolla tube is slightly crooked and 1.2 to 3.2 inches (3 to 8 cm) long.
Fruits: The fruits are spherical and .4 to .8 inch (1 to 2 cm) thick with a .8 inch (2 cm) long bill.
Occurrence: One finds this ornamental plant in parks. Often it is overgrown. Originally, it grew from Sikkim to Thailand and in the southern part of China.
Other names: English "Blue Trumpet Vine," "Sky Flower"; German "Großblütige Thunbergie," "Bengalische Trompete"; Spanish "Presidio de Amor."
Additional information: The Bengal clockvine grows very rapidly. With its offshoots, it can overgrow a large area within a short time. Although it can be kept outside in moderate climates, the herbaceous stems regularly freeze off. The African *Th. erecta,* with similar, but much darker violet flowers, is often cultivated as an ornamental bush. The *Thunbergia* species can be found in cultivation (see pages 188 and 190). Its appearance is entirely different.

Pink Mandevilla

Mandevilla splendens
Family: *Apocynaceae*, Dogbane

Most important features: All parts of the plant contain a milky sap. The leaves are set opposite. The flowers are large, pink, and shaped like a funnel.
Growth structure: The plant has thin, winding shoots and grows up to 13 feet (4 m) high. On rare occasions, it grows as a bush.
Leaves: The leaves are almost sessile, elliptical, and pointed clearly with emerging veins. They often have a weak heart-shaped bottom. The leaves are 3.2 to 8 inches (8 to 20 cm) long.
Flowers: The flowers grow individually or in small groups of five. They are 2 to 4.8 inches (5 to 12 cm) long. The corolla tube is wide. Often, it is outside; inside, it is either yellow or darker pink. It has five stamens, two separated ovaries, and two fleshy nectar glands.
Fruits: These are cigar-shaped. They grow in pairs, widely spread out. They are rarely seen in cultivation.
Occurrence: Often grown as an ornamental plant on a climbing frame, the pink mandevilla originally grew in the southeastern part of Brazil.
Other names: English "Red Riding Hood"; German "Dipladenie."
Additional information: One sees breeding specimens (*M. x amabilis*) in multiple colors from dark red to yellow in cultivation. Red riding hood is actually only the name of one kind. *M. sanderi*, whose leaves have petioles, is also given this name sometimes. The white blooming *Mandevilla laxa* is often found as ornamental plant. It is called Chilean jasmine although it is originally from Argentinia and not Chile. It can be easily differentiated from the real jasmine (see page 192) by its milky sap and its five stamens.

Chinese Honeysuckle

Quisqualis indica
Family: *Combretaceae*, Quisqualis plants

Most important features: The flowers have a long, very thin, stalklike tube. Originally, it is white, then pink, and finally, it is crimson. One side of the pistil adheres to the tube.
Growth structure: The plant climbs more than 33 feet (10 m) high. At first, it winds to the left. The older branches are anchored with thorns.
Leaves: The leaves are set opposite at blooming branches. Otherwise, they are alternating. They leaves are ovate to tapering. They are 2 to 7.6 inches (5 to 19 cm) long and 1 to 3.6 inches (2.5 to 9 cm) wide.
Flowers: Flowers grow on spikes .8 to 8 inches (2 to 20 cm) long that accumulate at the ends of the branches. The tube is 2 to 3.2 inches (5 to 8 cm) long. At the end, it has five small dentate sepals that grow from .2 to .6 inch (5 to 15 mm) long during the flowering season. The flowers have ten stamens. Five of them tower above the tube.
Fruits: The fruits are brown, dry, and ovate. They grow 1 to 1.6 inch (2.5 to 4 cm) long and .4 to .8 inch (1 to 2 cm) thick with five running wings.
Occurrence: This is a popular ornamental plant that often overgrows other plants. Originally, it was from Southeast Asia.
Other names: English "Drunken Sailors," "Rangoon Creeper"; German "Rangun-schlinger"; Portuguese "Arbusto Milagroso"; Spanish "Barbudo," "Santa Cecilia."
Additional information: Many specimens with filled flowers (see photo at right) are grown from the Chinese honeysuckle. This plant grows so rapidly that it often has to be cut back to protect the other plants in the garden. The thorns on its older branches grow out of petioles that remain after the plant sheds its leaf blades. Extracts of the leaves, roots, and unripe fruits are thought to be an efficient treatment for eelworms. The related, equally attractive hiccup nut (*Combretum* species) has a very short perianth and very long, bright red filaments.

Chinese Hat Plant

Holmskioldia sanguinea
Family: *Lamiaceae*, Labiate plants

Most important features: The flowers are brick red to orange (occasionally yellow). The calyx is spherical and is spread out like a Chinese straw hat with a thin, bent corolla tube.
Growth structure: At first bushy, the plant later climbs up to a height of 33 feet (10 m) with drooping, four-squared branches.
Leaves: The leaves are set opposite. They are tapering and ovate to elliptical. The leaves are 1.2 to 4.8 inches (3 to 12 cm) long and .6 to 3.2 inches (1.5 to 8 cm) wide, often with a serrated margin.
Flowers: The flowers grow in groups in the axils. The calyx is .8 to 1.2 inch (2 to 3 cm) wide. In older flowers, the calyx has a striking net of veins. The corolla is .6 to 1 inch (1.5 to 2.5 cm) long. The margin has five lobes. The lower lobe is .2 inch (5 mm) long. The remaining ones are much smaller. The flowers have four stamens in two pairs of different lengths. They are slightly longer than the corolla tube.
Fruits: The fruits are brown, often warty, and shaped like a flattened sphere. Each fruit has four parts. Each part is .2 inch (5 mm) long and contains one individual pit. The fruits are entirely enclosed in the calyx.
Occurrence: This plant grows in all tropical areas. Originally, it spread from the southern slope of the Himalayas to Bangladesh.
Other names: English "Cup-and-Saucer Plant," "Mandarin's Hat," "Parasol Flower"; German "Chinesenhut."
Additional information: It is difficult to define the borderline between a bush and a vine with the Chinese hat plant. Sometimes, it can even be found as a tree with drooping branches. The broad calyxes, which give the plant its name, are spread out before the corolla unfolds. They remain intact far beyond the ripening of the fruits. At one time this species was assigned to the family of Vervains (*Verbenaceae*).

Lady's Slipper

Thunbergia mysorensis
Family: *Acanthaceae*, Acanthus plants

Most important features: The flowers are taller than they are wide. The corolla tube and base of the five curled corolla lobes are yellow; the tips of the lobes are often dark red. The four stamens and the bent pistil are located in the upper part of the flower.
Growth structure: Lady's slipper grows as a winding plant up to 33 feet (10 m) high.
Leaves: The leaves are set opposite. They are elliptical to tapering with three strong veins that start from the bottom. The leaves are 2.8 to 6 inches (7 to 15 cm) long and 1 to 3.6 inches (2.5 to 9 cm) wide.
Flowers: The flowers grow in drooping simple racemes up to 35 inches (90 cm) long. The petioles point downward, and the corolla tubes point upward. The calyx is enclosed in purple bracteoles that are .8 to 1 inch (2 to 2.5 cm) long. The corolla is up to 2.4 inches (6 cm) high and 1.6 inch (4 cm) wide.
Fruits: The fruits are spherical and about .4 inch (1 cm) long with a bill that is .8 inch (2 cm) long.
Occurrence: This striking creeper is not seen very frequently. It is occasionally grown on a trellis. It originated in the southern part of India.
Other names: English "Mysore Trumpet Vine"; German "Mysore-Schlinger."
Additional information: The lady's slipper needs lots of warmth. It cannot withstand temperatures lower than 50°F (10°C) without being damaged. Its flowers produce so much nectar that it often drips out of the tube. In India, nectar birds often hang onto the inflorescences. The hummingbirds of the American tropics hover in front of the flowers. Often, only two or three flowers are open at the same time, so the flowering season continues for a very long time. Because of the way the birds pollinate the plant, the flowers of the lady's slipper look completely different than other *Thunbergia* species (see pages 184 and 190); only *Th. coccinea* has similar flowers. However, they are red.

Black-Eyed Susan

Thunbergia alata
Family: *Acanthaceae*, Acanthus

Most important features: The leaves are set opposite. The bottom is heart-shaped to tapering. The flowers are yellow to orange with an almost black center.
Growth structure: The plant has herbaceous branches. It grows up to 13 feet (4 m) high.
Leaves: The leaves are ovate to triangular and tapering. On the bottom, they have five to seven veins. The continuous margin is dentate and the petioles are winged. The leaves are 1.4 to 6 inches (3.5 to 15 cm) long and 1 to 4.4 inches (2.5 to 11 cm) wide.
Flowers: The flowers grow individually or in small groups. The calyx has eleven to sixteen teeth. It is hidden between bracts that are .6 to 1 inch (1.5 to 2.5 cm) long. The corolla has five lobes and is 1.2 to 1.6 inch (3 to 4 cm) wide with a slightly bent tube that is .6 to .8 inch (1.5 to 2 cm) long.
Fruits: The fruits are spherical and about .4 inch (1 cm) thick. The bill is about .4 inch (1 cm) long. The fruits are enclosed in dry bracts.
Occurrence: The black-eyed Susan is grown worldwide as an ornamental plant. It also grows wild as a weed. Originally, it came from tropical Africa.
Other names: English "Clock Vine"; German "Schwarzäugige Susanne"; Spanish "Ojitos Negros," "Principe Alberto."
Additional information: The black-eyed Susan is one of the few tropical plants that can be found in gardens in moderate climates, even if only as annual summer flowers. The plant loves sunny locations. It needs a great deal of water. Therefore, it thrives best in climates that are not too hot or too dry. It dies off at the first frost. Besides the wild yellow specimen, there are species with white and red flowers. Some of these do not have the typical dark center. More rarely, one can also find the very similar *Th. Gregorii.* It also orginated in Africa and has bright orange flowers. Some of the approximately ninety *Thunbergia* species are grown as ornamental plants (see pages 184 and 188).

Buttercups

Allamanda cathartica
Family: *Apocynaceae*, Dogbane

Most important features: All parts of the plant contain a milky sap. The leaves grow in whorls of three or four together. Occasionally, they are set opposite. The flowers are large, yellow, and shaped like a funnel. The fruits are thorny.
Growth structure: The plant climbs up to 50 feet (15 m) high. Sometimes, it is cut back to serve as ground cover or to grow as a bush.
Leaves: The leaves are tapered to obovate, stiff, and shiny. They are 2 to 6.4 inches (5 to 16 cm) long and .8 to 2.4 inches (2 to 6 cm) wide.
Flowers: The flowers grow individually or up to five together. The calyx tips are .2 to .6 inches (5 to 15 mm) long. The corolla tube is 1.6 to 3.2 inches (4 to 8 cm) long. At the bottom, it is very narrow and red. The upper half has reddish longitudinal stripes and is hairy at the height of the five stamens. The corolla has three to five lobes.
Fruits: The fruits are almost spherical to ovate and slightly flattened. They are 1.2 to 3.2 inches (3 to 8 cm) long and densely covered with thorns that are .2 to .8 inch (0.5 to 2 cm) long. The fruits open when ripe, revealing seeds .6 to 1 inch (1.5 to 2.5 cm) long with a membranous wing.
Occurrence: The plant is widely cultivated and often overgrown. It originated in the northeastern part of South America.
Other names: English "Golden Trumpet," "Yellow Allamanda"; German "Goldtrompete"; French "Liane à Lait," "Monette Jaune"; Spanish "Amanda," "Campana," "Copa de Oro," "San José."
Additional information: Although buttercups are very poisonous, the milky sap and bark are sometimes used as a laxative. In European greenhouses, one can often see breeding specimens with particularly large flowers; in Asia, the dwarflike, nonclimbing specimens are much more common. The closely related *A. blanchetii* with its light red violet flowers is seldom seen. The flowers are 2 inches (5 cm) long and twisted in the bud like a screw.

Bagflower

Clerodendrum thomsoniae
Family: *Lamiaceae*, Labiate plants

Most important features: The leaves are set opposite. The flowers are white with an emerging calyx and a crimson corolla.
Growth structure: This is a winding plant that grows up to 23 feet (7 m) high. When it grows as a bush, it is much smaller.
Leaves: The leaves are tapering and ovate with a continuous margin and distinct veins. The leaves are 2.4 to 7.2 inches (6 to 18 cm) long and 1.2 inch to 3.2 inches (3 to 8 cm) wide.
Flowers: The flowers grow in short, drooping panicles. The calyx is .8 to 1.2 inch (2 to 3 cm) long. It is ovate, tapering, and five-sided. Almost to the bottom, it has five parts. The corolla tube is barely longer than the calyx. The five corolla lobes are each about .4 inch (1 cm) long. The four stamens and the pistil stand about .8 inch (2 cm) above the flower.
Fruits: Until the fruits ripen, they are enclosed in the violet calyx. They are shiny and black. They break into four parts.
Occurrence: This is a popular plant for semi-shady locations. It is also used as an indoor plant. It originated in tropical West Africa.
Other names: English "Bleeding Heart Vine," "Broken Hearts," "Glory Bower"; German "Blutendes Herz."
Additional information: According to the legend, the bagflower grew out of the tear of an abandoned virgin. It is the most common of the approximately four hundred *Clerodendrum* species. Many of these are cultivated (see page 132). Until recently, they were listed as Vervain plants (*Verbenaceae*); however, this did not seem to be correct. *C. splendens*, which is similar to the bagflower and also comes from tropical Africa, can also grow as a bush or vine. However, its calyxes are smaller and at least partially streaked with red. *C. capitatum* always grows as a vine with white flowers in green calyxes.

Arabian Jasmine

Jasminum sambac
Family: *Oleaceae*, Olive tree

Most important features: The leaves are set opposite. The flowers are white with a narrow corolla tube and five to twelve calyx and corolla cusps. The flowers only have two stamens.
Growth structure: The branches of this bush climb or sometimes only droop. The plant grows up to 10 feet (3 m) high.
Leaves: The leaves are ovate to elliptical. The veins on the bottom side are clearly defined and emerging. The leaves are 1.2 to 4.8 inches (3 to 12 cm) long and .8 to 2.8 inches (2 to 7 cm) wide.
Flowers: The flowers grow in small, dense groups at the ends of branches. Only one or two are open at the same time. The corolla tube is .2 to .6 inch (5 to 15 mm) long. The corolla lobes are .2 to .6 inch (5 to 15 mm) long and .2 to .3 inch (5 to 8 mm) wide. The flowers have a wonderful aroma.
Fruits: The fruits are spherical and black. They are .2 to .3 inch (5 to 8 mm) long. The fruits seldom grow close together.
Occurrence: The Arabian jasmine is cultivated everywhere in the tropics. It probably originated in India.
Other names: German "Arabischer Jasmin"; Spanish "Jazmín de Arabia."
Additional information: The Arabian jasmine owes its name to the fact that Arab merchants brought its pleasant-smelling flowers to Europe in the Middle Ages. Today, the flowers are used to add aroma to jasmine tea. The oil is used in making perfume. For Hindus and Buddhists, the flowers are a symbol of purity and are a popular offering in ceremonies. In cultivation, one can often find specimens with filled flowers. These do not have stamens and can only be reproduced with cuttings. Many others of the more than two hundred *Jasminum* species serve as ornamental plants, and not only in the tropics. *J. polyanthum*, for example, with its pinnately divided leaves and corolla tube often streaked red, is frequently seen as an indoor plant. The winter jasmine (*J. nudiflorum*) with its yellow flowers decorates gardens in the early spring.

Elephant Creeper

Argyreia nervosa
Family: *Convolvulaceae*, Bindweed

Most important features: The leaves are large and heart-shaped. The bottom side of the leaves is hairy. The flowers have a white calyx and a red violet corolla.
Growth structure: The elephant creeper grows as a winding plant, reaching up to 33 feet (10 m) high. The younger shoots are herbaceous with dense, light hair; the older ones are woody.
Leaves: The leaves are alternating with slightly immersed veins on the upper side. On the bottom side, the veins are firmly defined and emerging. The leaves are 4 to 12 inches (10 to 30 cm) long and 3.2 to 10 inches (8 to 25 cm) wide.
Flowers: Five flowers grow together on a long petiole. The bud is enclosed in a pale green bract that is 1.4 to 2 inches (3.5 to 5 cm) long. The sepals are ovate and .4 to .8 inch (1 to 2 cm) long. The corolla is funnel-shaped to tubular and about 2.4 inches (6 cm) long. On the outside, it has five hairy segments. The stamens and pistil are deep in the tube.
Fruits: The fruits are spherical, yellowish brown, and pointed. They are .4 to .8 inch (1 to 2 cm) long and surrounded by slightly larger sepals.
Occurrence: The elephant creeper is widely cultivated. Occasionally, it also grows wild. It was originally from India.
Other names: English "Baby Wood-Rose," "Elephant Ear Vine," "Silver Morning Glory"; German "Elefantenwinde."
Additional information: Within a very short time, the elephant creeper overgrows everything it can cling to. Its dried out fruits have enlarged sepals that are notably reddish on the inside. These can sometimes be found in dried flower arrangements. In Asia, one often sees some of the approximately ninety *Argyreia* species, yet this is the only one that is cultivated everywhere in the tropics.

Cairo Morning Glory

Ipomoea cairica
Family: *Convolvulaceae*, Bindweed

Most important features: The leaves are shaped like a hand. They are divided into five to seven lobes (almost palmately divided). The flowers are funnel-shaped and light red violet with a dark throat or with a white margin that is seldom entirely white.
Growth structure: The plant grows with herbaceous shoots up to 16 feet (5 m) high. It winds and, more rarely, crawls.
Leaves: The leaves are alternating. The lobes are tapering to ovate. The lower lobes are often divided once or twice more. The leaves are 1.2 to 2.4 inches (3 to 6 cm) long and .4 to .8 inch (1 to 2 cm) wide.
Flowers: The flowers grow individually or in small groups. The sepals are .2 inch (5 mm) long. The corolla is 1.8 to 2.4 inches (4.5 to 6 cm) long.
Fruits: These are spherical capsules with two chambers. They are .4 to .6 inch (1 to 1.5 cm) long with four flaps that open. Inside are seeds that grow up to 1.6 inch (4 cm) long. The seeds are hairy along the edges.
Occurrence: One finds the Cairo morning glory everywhere in the tropics, including along roadsides and on fallow land. Its origin is uncertain.
Other names: English "Five-Leaf Morning Glory," "Mile-a-Minute," "Railway Creeper"; German "Kairowinde"; Spanish "Aurora."
Additional information: The Cairo morning glory grows so rapidly that it often turns into a problem. The same is also true for many other of the approximately 650 morning glory species (*Ipomoea spp.*, see also pages 200, 204, 226, and 270), most of which have simple heart-shaped leaves or leaves with three lobes. *I. purpurea* is a wild specimen with blue violet flowers and a light to white tube. It has reddish stripes in the middle of the corolla lobes. The closed buds of *I. tricolor* are red. Its flowers are blue with a lighter to white tube. There are specimens of both species in cultivation with lilac or pink stripes on a white background.

Giant Granadilla

Passiflora quadrangularis
Family: *Passifloraceae*, Passionflower

Most important features: The stem is firm, four-sided, and winged. The flowers are large and pink to red violet (occasionally white) with one very large corona consisting of many violet and white cross-striped threads.
Growth structure: The plant grows with climbing tendrils. It reaches a height of up to 130 feet (40 m). Usually, it is woody only at ground level.
Leaves: The leaves are alternating, wide ovate to elliptical, and tapering with a continuous margin at the bottom. The leaves are also spherical to heart-shaped at the bottom, and they may be dentate there, too. The leaves are 3.6 to 10 inches (9 to 25 cm) long and 2.4 to 7.2 inches (6 to 18 cm) wide with stipules that are .8 to 2 inches (2 to 5 cm) long.
Flowers: The individual flowers are 2.8 to 4.8 inches (7 to 12 cm) wide. The five calyxes and five corollas are very similar when viewed from above. In the middle, the column has five stamens, the ovary, and the three-headed stigmas.
Fruits: The fruits are ovate to tapering and yellow green. They are 4.8 to 15 inches (12 to 35 cm) long and 4 to 6 inches (10 to 15 cm) thick. The flesh is up to 1.6 inch (4 cm) thick. It is white and slightly sour. Inside are many gray seeds with glassy, whitish coats.
Occurrence: The giant granadilla is native to the Western Hemisphere. It is frequently planted there, but it is not seen very often in other places.
Other names: German "Königsgranadilla," "Riesengranadilla"; French "Barbadine"; Portuguese "Maracujá Mamão"; Spanish "Badea," "Granadilla Real," "Parcha," "Tumbo."
Additional information: Unlike passionflowers (see also page 208), the flesh of the giant granadilla is edible. It is very similar in appearance to the winged passionflower (*P. alata*), which has much smaller stipules. The sweet granadilla (*P. ligularis*) has large, cream white to pink flowers. Even with a similar corona, it is easy to differentiate because of its round shoots. Fruits are orange with light points and purple spots.

Banana Poka

Passiflora tripartita var. mollissima
Family: *Passifloraceae*, Passionflower

Most important features: Leaves have three lobes. Flowers have three bracts at the bottom of a long, green tube, five calyxes and five pink to pale lilac petals. In the midcolumn are the five stamens, the ovary, and the three-headed stigmas.
Growth structure: This plant grows with climbing tendrils up to 65 feet (20 m) long. The plant is round to slightly squared.
Leaves: The leaves are alternating and hairy with a dentate margin. Up to half of each leaf is deeply incised. The leaves are 2 to 6.8 inches (5 to 17 cm) long and 2.8 to 10 inches (7 to 25 cm) wide.
Flowers: The flowers grow individually. They droop and are 2 to 4 inches (5 to 10 cm) wide. The flower tube is 2 to 4.8 inches (5 to 12 cm) long and white at the end.
Fruits: The fruits are elliptical to long oval. They are green to pale yellow and softly hairy. The fruits grow up to 2 to 4.8 inches (5 to 12 cm) long and 1.2 to 1.8 inch (3 to 4.5 cm) thick. Inside are many glassy and orange seeds.
Occurrence: The banana poka is cultivated in the highlands. It originated in the Andes.
Other names: English "Banana Passion Fruit"; German "Curuba"; Spanish "Curuba de Castilla," "Tacso," "Tumbo."
Additional information: The banana poka is the most popular of the passionflowers that have long flower tubes. It grows so rapidly that it has become a serious threat to native flora on Hawaii. The very similar, but more firmly colored *P. mixta* is considered one parent of some red ornamental specimens. Some species with short flower tubes, such as *P. racemosa*, have even brighter red flowers. Passionflowers are available in all sizes and colors, from tiny to gigantic and from green to bright yellow (*P. citrina*) or red (compare also page 208). Many species have very striking leaves. For example, these may be bisected or horizontally oblong to the longitudinal axils. They may be speckled or even spotted, resembling butterfly eggs.

Costa Rican Nightshade

Solanum wendlandii
Family: *Solanaceae*, Solanum

Most important features: The shoots and leaves have curved thorns. The large flowers are lilac. They have five indistinct lobes with a yellow stamen cone in the middle.
Growth structure: These vines grow up to 20 feet (6 m) high. They also grow without support as ground cover.
Leaves: The leaves are alternating. At the thin shoots, the leaves are often simple or have three lobes. At other places, the leaves are pinnate with up to seven lobes. They grow up to 10 inches (25 cm) long and 6 inches (15 cm) wide with ovate to oval segments. The terminal segment is approximately 5.4 inches (13.5 cm) long and 4 inches (10 cm) wide.
Flowers: The flowers grow in panicles that are up to 8 inches (20 cm) long. The flowers are 1.4 to 2.4 inches (3.5 to 6 cm) long. Each lobe has a delicately pointed end. The stamen cone often has a lilac pointed end.
Fruits: The fruits, yellow to orange berries, are ovate and 1.4 to 3.2 inches (3.5 to 8 cm) long.
Occurrence: One seldom sees these plants growing wild. Usually, they grow in gardens. Originally, they were from Costa Rica.
Other names: English "Divorce Vine," "Giant Potato Creeper," "Grand Potato Vine," "Marriage Vine"; German "Costa-Rica-Nachtschatten"; Spanish "Elisa," "Jazmín Italiano," "Quixtán."
Additional information: The Costa Rican nightshade is probably the most striking of the climbing *Solanum* species (see also pages 104 and 150). Its fruits are eaten in Guatemala. The rest of the plant is considered poisonous, probably because of the danger of confusing it with other species. *S. seaforthianum* has no thorns. It also has smaller, star-shaped flowers with corolla lobes that are separated almost to the bottom. Its fruits are smaller and red. This was popular as an ornamental plant in many places, but it has turned out to be an aggressive weed. The jasmine nightshade (*S. Laxum*) is similar, but its flowers are bluish white and only about .8 inch (2 cm) long.

Bougainvillea

Bougainvillea spectabilis
Family: *Nyctaginaceae*, Four-o'clock

Most important features: Three flowers, each enclosed in three ovate bracts, grow on a petiole. The flowers are often crimson to pink violet.
Growth structure: This is a woody plant that reaches up to 80 feet (25 m) high. It climbs on trees as a vine. In Asia, it is often cut to grow as a bush. The plant grows with long, green shoots. These often have firm thorns in the axils.
Leaves: The leaves are alternating and ovate to tapering. They are 1.6 to 4 inches (4 to 10 cm) long and .8 to 2.4 inches (2 to 6 cm) wide with velvety hair on the bottom.
Flowers: The flowers grow in panicles that are in bracts. The panicles are 1 to 2.6 inches (2.5 to 6.5 cm) long and .6 to 1.6 inch (1.5 to 4 cm) wide. The flowers are often pale yellow. They are narrow and tubular. The flowers have five lobes and often have eight stamens in the tube.
Fruits: The fruits are pear-shaped and .4 to .6 inch (1 to 1.5 cm) long with five longitudinal grooves.
Occurrence: One finds bougainvillea in all tropical and subtropical areas. It was originally from Brazil.
Other names: German "Drillingsblume"; Spanish also "Flor de Verano," "Lustrosa," "Manto de Jesús," "Napoleón," "Pompilla," "Tres Marias," "Trinitaria."
Additional information: Bougainvillea is named after its discoverer, the French admiral, Louis Antoine, Comte de Bougainville. Today, crossbreeds of three very similar species are cultivated. They come in all colors from white to orange to red violet. Sometimes, they have a larger number of bracts. The leaves of the two other species, *B. peruviana* and *B. glabra*, have very little hair. *B. peruviana* has smaller leaves that may even be bare. *B. glabra* can withstand frost up to 20°F (-7°C).

Cardinal Climber

Ipomoea quamoclit
Family: *Convolvulaceae*, Bindweed

Most important features: The leaves are pinnatifid (deeply indented) and divided into many narrow lobes. The flowers are bright red with a narrow tube; they form a five-sided star.
Growth structure: This plant grows with thin, herbaceous shoots. It winds up to 10 feet (3 m) high and rarely crawls.
Leaves: The leaves are alternating and ovate to tapering. They are .6 to 4 inches (1.5 to 10 cm) long and .4 to 2.4 inches (1 to 6 cm) wide with seventeen to forty-one lobes. The lowest lobes are often divided once more.
Flowers: The flowers grow individually or a few grow together. The tapering sepals are .2 to .3 inch (5 to 8 mm) long. The corolla tube is .8 to 1.4 inch (2 to 3.5 cm) long. The stamens and pistil rise above the tube.
Fruits: The fruits are tapering, ovate capsules about .4 inch (1 cm) long. They have four flaps that open, revealing the four large, dark seeds.
Occurrence: The cardinal climber is widely spread as an ornamental plant. Often, it also grows wild. It originated in the Western Hemisphere.
Other names: English "Cypress Vine," "Cupid Flower," "Red Jasmine," "Star Glory"; German "Kardinalswinde"; French "Cheveux de Venus," "Liane Rouge"; Portuguese "Corda de Viola"; Spanish "Cabello de Angel," "Cundeamor," "Estrella del Sol," "Regadero."
Additional information: In warm, protected places, the cardinal climber grows so rapidly that it can even be cultivated in moderate climates as an annual summer flower. It differs from most of the other *Ipomoea* species (see pages 194, 226, and 270) because it doesn't have funnel-shaped flowers. Instead, it has a long, narrow tube. The long tube indicates that it is pollinated by butterflies. The closely related *I. alba*, which is 2.8 to 6 inches (7 to 15 cm) long with white flowers, is pollinated by nighthawks, birds with a long rostrum.

Bride's Tears

Antigonon leptopus
Family: *Polygonaceae*, Knotgrass

Most important features: On the upper side of the leaves, the veins are immersed; on the bottom side, the veins are emerging. The flowers are pink (occasionally white) with three large, outer bract whorls and two smaller, inner whorls.
Growth structure: Bride's tears climbs up to 40 feet (12 m) high with herbaceous shoots from a woody base.
Leaves: The leaves are alternating, triangular to heart-shaped, and acute. The margins are wavy and often lightly indented. The leaves are 1.2 to 6 inches (3 to 15 cm) long and .8 to 4.8 inches (2 to 12 cm) wide.
Flowers: The flowers grow in simple racemes or panicles. The tips are turned into tendrils. The outer bract whorls are heart-shaped and up to .4 inch (1 cm) long and .3 inch (8 mm) wide. The inner ones are elliptical and narrower. The eight stamens are connected halfway up. Often, there is a single spur between the filaments.
Fruits: The fruits are brown nuts. They have three edges and are pyramid-shaped. The nuts are about .4 inch (1 cm) long and enclosed in three outer bract whorls that are enlarged by up to 1 inch (2.5 cm).
Occurrence: The plant is widely cultivated and often grows wild. Originally, it was from Mexico.
Other names: English "Chain of Love," "Confederate Vine," "Coral Vine," "Honolulu Creeper," "Mexican Creeper," "Pink Vine"; German "Mexikanischer Knöterich"; French "Liane Corail"; Spanish "Cadena de Amor," "Colación," "Coralillo," "Coralita," "Flor de San Miguel," "San Diego."
Additional information: Bride's tears grows very rapidly and can cover fences and other plants with a dense latticework of shoots. The shoots flower almost all year long. The plant needs to be cut regularly to control it. In Thailand, its young leaves and flowers are fried and eaten or are eaten covered with egg and bread crumbs. The root tubercle, which is rich in starch, is also supposed to be edible.

Red Bauhinia

Bauhinia galpinii
Family: *Caesalpiniaceae*, Carob plants

Most important features: The leaves are spherical and have two lobes. The flowers are large and attached to petioles. The petals are bright red.
Growth structure: The red bauhinia is a climbing plant or a bushy tree with drooping branches. It seldom exceeds 16 feet (5 m). As a tree, it has a wide top.
Leaves: The leaves are alternating and up to 4.8 inches (12 cm) long. At the bottom, they are spherical to heart-shaped. At the tip, they have wide grooves that extend approximately 30 percent of the length.
Flowers: The flowers grow individually or in small flowering panicles at the ends of the branches. The petals are salmon to crimson in color, paddle-like, and ordered on one petiole. The leaf blade is spherical to heart-shaped. The flowers are 2.4 to 3.2 inches (6 to 8 cm) wide.
Fruits: The fruits are up to 4 inches (10 cm) long and .8 inch (2 cm) wide. They are flattened. When they are ripe, the fruits turn dark brown and burst into two parts.
Occurrence: The red bauhinia is often used as an espalier. It is found in tropical and subtropical gardens. Originally, the plant was from the southeastern part of Africa.
Other names: English "Pride of De Kaap"; German "Rote Bauhinie."
Additional information: The red bauhinia is a very robust plant. During a drought or when exposed to cold temperatures, it loses its leaves, but it quickly sprouts again. In contrast to the red bauhinia, many of the numerous vines among the approximately three hundred *Bauhinia* species (compare with pages 98 and 172) have creepers in addition to leaves. One of the most frequently occurring creeping species is the Southeast Asian *B. corymbosa* (phanera). It has small, almost entirely incised leaves and bunches of pale pink flowers. Each flower has five firm stamens and five petals. Three of the petals point upward, and two of the petals point downward. Many climbing species can contract their branches in waves or elastic springs, then grow in width.

Climbing Lily

Gloriosa superba
Family: *Colchicaceae*, Meadow saffron

Most important features: The tips of the leaves lengthen into tendrils. The flowers are large and often red and yellow. The pistil is bent sideways.
Growth structure: The plant climbs 4.8 to 20 feet (1.5 to 6 m) high with herbaceous shoots that grow from a tuberous root system.
Leaves: Most of the leaves are alternating. However, in parts they are also set opposite or grow in whorls of three, sessile. The leaves are 3.2 to 10 inches (8 to 25 cm) long (without the tendril) and .4 to 1.8 inch (1 to 4.5 cm) wide. They are ovate to almost grasslike. The leaves are very delicately pointed with veins running parallel to the midrib.
Flowers: These grow individually on long petioles and are turned toward the ground. The petals, 2 to 3.6 inches (5 to 9 cm) long and .4 to 1.2 inch (1 to 3 cm) wide, are curled toward the top. Often they are yellow at the bottom and red at the tip, only rarely one color. During the flowering season, they become darker. The stamens spread out sideways and are 1 to 2 inches (2.5 to 5 cm) long. The green ovary has three distinct parts.
Fruits: The fruits are leathery. They are 1.6 to 4 inches (4 to 10 cm) long and .6 to 1.4 inch (1.5 to 3.5 cm) thick. When they are ripe, three flaps open, revealing bright red seeds.
Occurrence: The plant has spread everywhere in the tropics.
Other names: English "Glory Lily," "Flame Lily," "Malabar Lily," "Superb Lily"; German 'Ruhmeslilie," "Flammenlilie," "Prachtlilie," "Tigerklaue"; French "Lis de Malabar"; Spanish "Gorra de Turco," "Lirio Gloriosa," "Lirio de Malabar."
Additional information: The climbing lily grows wild only in dry areas, and the parts above the ground die off. It contains the same poison as meadow saffron (*Colchicum autumnale*), especially in the tubercle. As an ornamental plant, the species, *Rothschildiana* is particularly popular. It has wavy petals that are green at the bottom, yellow in the middle, and red toward the end of the petal.

Exotic Love

Ipomoea lobata
Family: *Convolvulaceae*, Bindweed

Most important features: Most of the leaves have three lobes. The flowers are oblong, urn-shaped, five-sided, and slightly bent. They are red in the bud, but change to yellow and almost white as they unfold.
Growth structure: The plant grows with thin, herbaceous shoots, winding up to 20 feet (6 m) high.
Leaves: The leaves are alternating and heart-shaped with up to five lobes. They are 2.4 to 6 inches (6 to 15 cm) long and wide. The lobes are tapering. The two lobes on the sides are often coarsely dentate.
Flowers: The flowers grow in inflorescences up to 16 inches (40 cm) long. Sometimes they are branched, erect, and simple racemes. The sepals are about .2 inch (0.5 cm) long with a pointed end. The corolla is .8 to 1.2 inch (2 to 3 cm) long. The flowers that are just above the ground are the widest. They have a narrow nozzle and five tiny, red cusps. The stamens and pistil extend far beyond the tube.
Fruits: The fruits are ovate and about .2 to .3 inch (5 to 8 mm) long. When they are ripe, the four flaps open, releasing the seeds.
Occurrence: One finds exotic love grown as an ornamental plant all over the world. In areas with moderate temperatures, it is grown as an annual summer flower. It originated in tropical and subtropical areas of the Western Hemisphere.
Other names: English "Firecracker Vine," "Spanish Flag"; German "Spanische Flagge"; Spanish "Bandera Española."
Additional information: Exotic love is named for its red and yellow buds, which often all face one side of the inflorescence and stand out almost horizontally, as if from a flagpole. Because of its tubular flowers, it differs markedly from other species of *Ipomoea* (see pages 194, 200, 226, and 270). In fact, it is often put into its own genus. In the Western Hemisphere, it is considered the best plant to attract hummingbirds.

Balsam Pear

Momordica charantia
Family: *Cucurbitaceae*, Gourd

Most important features: Tendrils are unbranched. Hand-shaped leaves have five to seven lobes. Flowers are unisexual. Yellow petals have distinct veins. The fruits tend to be warty.
Growth structure: The balsam pear grows up to 16 feet (5 m) long with climbing or crawling herbaceous shoots.
Leaves: The leaves are alternating, 1.2 to 6.8 inches (3 to 17 cm) long and wide. The lobes are obovate to rhombic with a narrow bottom. The margin is coarsely dentate to pinnately lobed.
Flowers: The flowers grow individually with a spherical branch at the petiole. Male flowers have petals that are .6 to .8 inch (1.5 to 2 cm) long and three stamens. Female flowers are smaller with three two-lobed stigmatic branches. The ovary is spindle-shaped, and .4 to 1.2 inch (1 to 3 cm) long. Female flowers are warty with soft thorns.
Fruits: The fruits have various shapes, but often they are pointed at the ends. They are1.4 to 12 inches (3.5 to 30 cm) long and .8 to 3.2 inches (2 to 8 cm) thick with eight to ten longitudinal ribs. The fruits are densely warty. When they are ripe, they are orange. When the three flaps open, the seeds hang out with red hulls.
Occurrence: One finds balsam pears in all tropical areas, both as useful plants and as wild flora. The place of origin is uncertain.
Other names: English "Bitter Gourd"; German "Bittergurke," "Balsambirne"; French "Concombre Africain," "Margose"; Spanish "Bálsamo," "Cundeamor."
Additional information: People in all tropical areas eat the young shoots and leaves of the balsam pear. Uncooked, the fruits are very bitter and often used for medicinal purposes. Several of the many cultivated forms can be made edible by marinating them in saltwater for a long period of time and boiling them thoroughly afterward. The closely related southern balsam pear (*M. balsamina*) differs from the balsam pear because the lobes of its leaves are not as deep.

Cup of Gold

Solandra maxima
Family: *Solanaceae*, Solanum

Most important features: The leaves are alternating. The flowers are very large with a tube that is narrow at the bottom and bell-shaped toward the top. The flowers have five lobes that are yellowish. Each lobe has one violet to brown stripe that extends from the middle of the lobe to deep into the corolla tube.
Growth structure: The plant climbs up to 40 feet (12 m) tall. When cultivated, it is often cut back to a bush with drooping branches.
Leaves: The leathery leaves are wide and elliptical to tapering. Often, they are pointed. They are 2 to 7.2 inches (5 to 18 cm) long and .8 to 3.6 inches (2 to 9 cm) wide. On the upper side, they are shiny.
Flowers: The flowers usually grow individually. The calyx is 2 to 3.2 inches (5 to 8 cm) long and five-sided with three to five cusps. The corolla is 6 to 9.6 inches (15 to 24 cm) long with large, rounded lobes and five stamens. The corolla is 3.2 to 6 inches (8 to 15 cm) wide. The flowers are cream white to light yellow; later, they are darker. Finally, they are ocher to orange.
Fruits: The fruits are spherical berries that are up to 2.8 inches (7 cm) long with the remains of the calyx.
Occurrence: The cup of gold is popular as an ornamental plant, especially for use along fences. It originated in Mexico.
Other names: English "Golden Chalice Vine," "Trumpet Plant"; German "Goldbecher"; Spanish "Bolsa de Judas," "Copa de Oro," "Gorro de Napoleón."
Additional information: Several species of the *Solandra* genus are so similar that different botanical guides define them differently or even consider them to be the same. Because of its strong growth, cup of gold is not a good choice for small gardens. Its buds are filled with water before they open. This water is supposed to help treat conjunctivitis. However, the plants themselves are poisonous and are also used to produce intoxicating drugs. With the naked eye, one can watch the flowers unfold every evening. They spread a strong fragrance that attracts bats for pollination.

Club Gourd

Trichosanthes cucumerina
Family: *Cucurbitaceae*, Gourd

Most important features: The leaves have hand-shaped lobes. The flowers are white with a long tube. The five corolla cusps have long fringes that are often curled.
Growth structure: The plant crawls or climbs, reaching a height of up to 20 feet (6 m) with branched off tendrils.
Leaves: The leaves have three to seven lobes that are scalloped or dentate and hairy. The leaves are 2 to 10 inches (5 to 25 cm) long and just as wide.
Flowers: The flowers are unisexual. The male ones grow in simple racemes; the female ones grow individually. The flowers are 1.2 to 1.8 inch (3 to 4.5 cm) wide. The floral tube is .6 to 1 inch (1.5 to 2.5 cm) long. At its end, there are five short calyx cusps and five corolla cusps. Spread flat, they would be .2 to .6 inch (0.5 to 1.5 cm) long.
Fruits: The shape of the fruits is very variable. Wild species are usually spindle-shaped. The fruits are up to 2.4 inches (6 cm) long. They are often orange red. Cultivated forms may grow up to 7 feet (2 m) long; however, they are only 2 to 2.4 inches (5 to 6 cm) thick. Often they are twisted snakelike and striped in different colors.
Occurrence: One finds the club gourd as an ornamental plant in all tropical areas. In West Africa and Southeast Asia, it is also cultivated as a useful plant. Originally, the plants grew from Pakistan to the Philippines and up to the northern part of Australia.
Other names: English "Serpent Gourd," "Snake Gourd"; German "Haarblume," "Schlangengurke."
Additional information: Club gourds are short-lived, rapidly growing flowers. They can also be grown in moderate latitudes as annuals. They are most beautiful in the evening and early morning to attract the nighthawks that pollinate them. During the day, the gracious flowers dry out easily. The fruits of the cultivated specimen, *var. anguina*, are called snake gourds. They are harvested and prepared the same way as one would prepare zucchini. However, club gourds have little taste. The very young shoots and leaves are sometimes used as vegetables.

Passion Fruit

Passiflora edulis
Family: *Passifloraceae*, Passionflower

Most important features: The bare leaves have three deep lobes. The flowers are slightly greenish white. Above the flowers is a corona consisting of many wavy threads. The threads are violet at the bottom and white toward the tip.
Growth structure: The passion fruit climbs with tendrils. The branches reach up to 50 feet (15 m) high. The plant is woody at ground level.
Leaves: The leaves are alternating and elliptical to tapering. They are 2 to 8.8 inches (5 to 22 cm) long and wide. The margin has dentate lobes.
Flowers: The flowers grow individually at the bottom of three ovate bracts that are .6 to 1 inch (1.5 to 2.5 cm) long. The flowers have five calyxes and five corollas. Seen from above, these are very similar. They are 2 to 3.2 inches (5 to 8 cm) wide. The ovary and three-headed stigmas and five yellow stamens are in the middle column.
Fruits: The fruits are spherical to ovate with a tough peel that is dark purple or yellow. The fruits are 1.6 to 4 inches (4 to 10 cm) long. Inside are many dark seeds, enclosed in a glassy, yellow orange coat.
Occurrence: One finds the passion fruit cultivated and growing wild. Originally, it was from South America.
Other names: English "Purple Granadilla"; German "Maracuja," "Pupurgranadilla"; French "Grenadille"; Portuguese "Maracujá"; Spanish "Maracuyá."
Additional information: Some people enjoy drinking the juice pressed from the coat of the seeds of the passion fruit. Many of the approximately 430 Passionflower species (see also page 196) are cultivated, and some are quite similar to the passion fruit. As indoor plants, one finds crossbreeds of the blue passionflower (*P. caerulea*), which has five to seven lobes per leaf and wide stipules. The water lemon or golden bellapple (*P. laurifolia*) has undivided leaves, and its corona is wider than the corolla itself. The fetid passionflower (*P. foetida*), frequently growing as a weed, is easily recognizable because of the delicately ripped hull around the flowers.

Pelican Flower

Aristolochia gigantea
Family: *Aristolochiaceae*, Aristolochia

Most important features: The flowers are very large with a hanging, balloon-shaped, widened base. They have a narrow tube that is bent upward. This leads into a spread out, purple edge in the front with light veins.
Growth structure: The pelican flower winds up to 33 feet (10 m) high. It is a woody vine with a deeply torn, corky bark. The young shoots are herbaceous and bare.
Leaves: The leaves are alternating, ovate to heart-shaped, and tapering. They are 4 to 8 inches (10 to 20 cm) long and 2.8 to 4.8 inches (7 to 12 cm) wide. On the bottom, they have dense, white hair.
Flowers: The flowers grow individually or a few together at older shoots. At the base, they are up to 3.2 inches (8 cm) long and 1.2 inch (3 cm) wide. The edge is 6 to 8 inches (15 to 20 cm) long and 4.8 to 6.4 inches (12 to 16 cm) wide. They are deeply scalloped at the bottom. The ovary is below, and the six stamens and stigmatic branches create a head-shaped formation at the bottom of the flower.
Fruits: The fruits are cylindrical capsules 3.2 to 5.2 inches (8 to 13 cm) long and .4 to 1.2 inch (1 to 3 cm) thick. The numerous seeds burst open from the bottom.
Occurrence: As an ornamental plant, the pelican flower is widely spread. Originally, it grew from Panama to the Amazon.
Other name: German "Riesen-Pfeifenblume."
Additional information: The flowers of the pelican flower look like a carcass, and they smell like one, too. Their odor attracts flies, which crawl into the tube. A hair keeps them inside for long enough that they are completely covered with pollen. Many of the approximately 120 *Aristolochia* species also set such traps. However, most of the other species have smaller flowers, like the native birthwort (*A. clematitis*). The large-flowered pelican flower (*A. grandiflora*) is often confused with the pelican flower. However, it has 4.8 inches (12 cm) to more than 13 feet (4 m) at the lower end of the flower margin.

Ceriman

Monstera deliciosa
Family: *Araceae*

Most important features: Ceriman has thick shoots with numerous hanging aerial roots. The leaves are often gigantic with firm lateral veins and deep lobes with holes.
Growth structure: The plant grows by climbing, often more than 33 feet (10 m) high. Its shoots are up to 2.8 inches (7 cm) thick. Sometimes, the plant grows on other trees, sending aerial roots to the ground.
Leaves: The leaves are alternating with firm petioles. The flanks of the petioles cling to the shoot. The leaves are leathery and dark green with light veins. Young leaves are small, heart-shaped, and undivided. Older leaves grow up to 3 feet (1 m) long
Flowers: The flowers are tiny and grow on erect, cream-colored cobs that are enclosed in a light bract. The cobs are 12 inches (30 cm) long and 2 inches (5 cm) thick.
Fruits: The fruits are berries that grow up to .4 inch (1 cm) long. These are crammed together on the cob, which is 3.6 inches (9 cm) thick in the flowering season. When ripe, the cob is streaked with cream-white or violet.
Occurrence: Ceriman grows on trees and in houses, often as a potted plant. It originated in Central America.
Other names: English "Mexican Breadfruit," "Split Leaf Philodendron," "Swiss Cheese Plant"; German "Fensterblatt"; Portuguese "Banana de Macaco"; Spanish "Harpón," "Hojadillo," "Piñanona."
Additional information: The ceriman is appreciated as an ornamental plant and for its fruits. The fruits smell and taste like a mixture of pineapple and banana. However, one must be careful because they often contain oxalic acid and other poisonous substances that can irritate the mucous membranes and produce other side effects. People often create robes and baskets from the aerial roots. The young form of the plant is reminiscent of climbing *Philodendron* and is known as *Ph. pertusum.*

Devil's Ivy

Epipremnum pinnatum 'Aureum'
Family: *Araceae*

Most important features: Devil's ivy is a root climber with long petiolates. The leaves are ovate to heart-shaped. They have yellow stripes and spots. On older plants, the leaves can be very large and have holes and deep cuts on the side.
Growth structure: This is a climbing plant that grows up to 50 feet (15 m) high. The fleshy shoot has aerial roots.
Leaves: The leaves are alternating with firm petioles that cling to the shoot with membranous flanks. On young plants, the leaves are undivided and up to 6 inches (15 cm) long; on large plants, they are up to 32 inches (80 cm) long and 24 inches (60 cm) wide.
Flowers: Hundreds of tiny flowers grow on cobs approximately 6 inches (15 cm) long. The cob is surrounded by a pale green to yellowish whorl of bracts.
Fruits: The fruits are small, yellowish berries densely crammed together on the cob.
Occurrence: Devil's ivy grows anywhere in the tropics. One finds it on house walls and tree trunks. Originally, it was from the Solomon Islands.
Other names: English "Golden Pothos," "Marble Queen"; German "Efeutute," Goldranke," "Buntes Herzblatt."
Additional information: We know devil's ivy as a modest ornamental plant that survives even in the darkest corners of our offices and apartments where it never grows beyond its youthful shape. However, in the tropics, this becomes a very large plant, which frequently catches a visitor's eye. Only *Aureum*, with its yellow, spotted leaves, is cultivated. The wild specimen has green leaves and can only be differentiated from relatives in the *Monstera* and *Philodendron* genera because of small differences in the flowers.

Bluewings

Torenia fournieri
Family: *Scrophulariaceae*, Brownwort

Most important features: This is an herbaceous plant with a four-sided stem. The flowers are light violet. The two corolla lobes on the side are dark violet; the lower one has a dark violet margin and a yellow spot at the throat.
Growth structure: Bluewings grows as an erect or slightly crawling plant that is often richly branched. It grows up to 6 to 15 inches (15 to 35 cm) high or up to 24 inches (60 cm) long.
Leaves: The leaves are set opposite. They are ovate, tapering, and light green with a serrate margin. The leaves are 1 to 2 inches (2.5 to 5 cm) long and .4 to 1.2 inch (1 to 3 cm) wide.
Flowers: The flowers grow individually or in pairs in the axils and as small groups at the ends of the stems. The calyx has five lobes and is .6 to .8 inch (1.5 to 2 cm) long with wing-shaped edges. The corolla is 1 to 1.4 inch (2.5 to 3.5 cm) long with a slightly bent tube and four lobes. The upper lobe is slightly wider than the others and slightly scalloped in the middle. Among the lobes are two pairs of stamens, which are not of equal length, and a stigma with two lobes.
Fruits: The fruits are elliptical capsules with two chambers. They are about .4 inch (1 cm) long and .2 inch (5 mm) thick and enclosed in the calyx.
Occurrence: One finds bluewings as an ornamental plant and also as a weed all over the world. Originally, it was from Southeast Asia.
Other names: English "Florida Pansy," "Wishbone Plant"; German "Schnappmäulchen"; Spanish "Lazo de Amor."
Additional information: Bluewings is an attractive, yet short-lived plant. Because it flowers only a few weeks after it germinates, it can be grown as a summer flower. There are varieties with blue, pink, yellowish, or white flowers as well as hanging specimens that grow well on balconies. While bluewings prefer rather shady locations, with sufficient water, they can also thrive in full sun.

Bachelor's Button

Gomphrena globosa
Family: *Amaranthaceae*, Amaranth

Most important features: The plant is delicately hairy and often streaked with red. The leaves are set opposite. The inflorescences are spherical to ovate and pink to red (occasionally white).
Growth structure: This plant grows erect or climbs. It is an annual that reaches 6 to 24 inches (15 to 60 cm).
Leaves: The leaves are ovate to tapering. At the bottom, they gradually change into short petioles. The leaves are 1 to 6 inches (2.5 to 15 cm) long and .8 to 2.4 inches (2 to 6 cm) wide,
Flowers: The inflorescences are .6 to 1 inch (1.5 to 2.5 cm) long. Each flower is .2 inch (5 mm) long and has a leaf and two narrow, stiff, reddish bracts that are .3 to .6 inch (8 to 15 mm) long. The leaves are dentate with a pointed end. The flowers are yellowish and hairy with five stamens. The cusps are between the anthers.
Fruits: The fruits are spherical to ovate, reddish, dry, and hidden between the bracts. They are less than .2 inch (5 mm) long.
Occurrence: Bachelor's button grows in gardens and wastelands in all tropical areas. Originally, it grew in the Western Hemisphere.
Other names: English "Globe Amaranth"; German "Kugelamaranth"; Spanish "Amarantina," "Amor Seco."
Additional information: The bachelor's button is often used for flower arrangements because its inflorescences keep their shape and color for a very long time, even in dried bouquets. The flowers themselves wilt, but they are insignificant compared with the colorful bracts. One can find specimens of all colors, from white to orange up to deep violet. Because the bachelor's button flowers a few weeks after it germinates, it can be planted as a summer flower. In tropical ornamental gardens, it is replaced every three months. Otherwise, it tends to become unattractive.

Four-o'clock

Mirabilis jalapa
Family: *Nyctaginaceae*, Four-o'clock

Most important features: The leaves are set opposite. The flowers grow in groups of five. They are often reddish with a long tube and a funnel shape. At first, the stamens and pistil are curled in the tube.
Growth structure: This is a richly branched herb that grows 20 to 60 inches (50 to 150 cm) high. The stipules are often slightly reddish. The root is thick and beet-shaped.
Leaves: The leaves are ovate, thin, tapering, and spherical to almost heart-shaped at the bottom. They are 1.2 to 4 inches (3 to 10 cm) long and .8 to 2 inches (2 to 5 cm) wide.
Flowers: The flowers grow in dense groups at the end of the shoot, but they don't always all open at the same time. They are often crimson to red violet, but they are also white, yellow, orange, or striped. The tube is 1.2 to 2 inches (3 to 5 cm) long. Above the thick ovary, the tube is very thin at first. The flower is 1 to 1.4 inch (2.5 to 3.5 cm) wide. The stamens and pistil protrude slightly above the tube.
Fruits: The fruits are ovate, wrinkly, black, and enclosed in a cuplike hull. They are .3 inch (8 mm) long,
Occurrence: Four-o'clock is an ornamental plant and weed found in many tropical and subtropical areas. It originated in Mexico and the southern part of the United States.
Other names: English "Marvel of Peru"; German "Wunderblume"; French "Belle de Nuit"; Spanish "Buenas Tardes."
Additional information: The four-o'clock lives for several years; however, it can also be cultivated as an annual herb. It owes its name to the fact that the flowers open late in the afternoon and close again toward morning. What looks like a calyx is actually a perianth of the bract. The roots, called "false jalap," are a strong laxative, and the flowers are used for food coloring in Asia.

Madagascar Periwinkle

Catharanthus roseus
Family: *Apocynaceae*, Dogbane

Most important features: The leaves are set opposite with a light midrib. The regular flowers have five petals, which are pink (occasionally white) and darker in the middle. They have a narrow, thickened tube below the margin of the corolla.
Growth structure: An herbaceous plant or a subshrub, the plant grows 12 to 24 inches (30 to 60 cm) high with many fleshy branches.
Leaves: The leaves are tapering, shiny, and often spherical at the tip. They are widest above the middle. The leaves are 1 to 4 inches (2.5 to 10 cm) long and .4 to 1.2 inch (1 to 3 cm) wide.
Flowers: The flowers grow individually or in pairs. They are 1.2 to 2 inches (3 to 5 cm) wide. The tube is .8 to 1.2 inch (2 to 3 cm) long. The corolla lobes are twisted in the bud. The five stamens grow in the bulge of the tube.
Fruits: Two fruits grow from each flower. They are green, .4 to 1.4 inch (1 to 3.5 cm) long, and less than .2 inch (5 mm) thick. When they are ripe, they open lengthwise, exposing many black seeds.
Occurrence: One finds the Madagascar periwinkle everywhere in the tropics. It originated in Madagascar in sandy locations near the coast.
Other names: English "Old Maid," "Rosy Periwinkle"; German "Catharanthe," "Madagassisches Immergrün"; Spanish "Chula," "Clavelina," "Pervinca de Madagascar," "Vicaria."
Additional information: The Madagascar periwinkle is also popular as an indoor plant. It almost always has flowers. As a result, it is sometimes sold as "busy Lizzie," although this name actually belongs to *Impatiens walleriana* (see page 216), the flowers of which are rich in color and have a long spur. Like most representatives of its family, the Madagascar periwinkle is poisonous, yet several of its alkaloids have proven to be effective against some kinds of cancers. However, the substances are contained in such small amounts that more than 1,100 pounds (500 kg) of plants are necessary in order to produce a fraction of an ounce (gm) of the substance.

New Guinea Impatiens

Impatiens hawkeri
Family: *Balsaminaceae*, Touch-me-not

Most important features: The leaves are set opposite or in whorls of three. The flowers have two small sepals on the side. One is larger than the other. The lower sepal has a long, crooked spur as well as five bright red petals.
Growth structure: This is an herb with juicy branches. It grows 8 to 24 inches (20 to 60 cm) high.
Leaves: The leaves are tapered at both ends. They are dark green to olive green or reddish. The midrib is often red, and the adjacent leaf blade may be light yellowish. The leaves are 2 to 6 inches (5 to 15 cm) long and .8 to 2.8 inches (2 to 7 cm) wide with a delicately dentate margin.
Flowers: The flowers grow individually or a few grow together. The petals are slightly scalloped at the tip. The stamens are connected to a red to yellowish bonnet that is rapidly shed. After the bonnet is shed, a green ovary is recognizable in the center of the flower. The flowers are 1.6 to 2.8 inches (4 to 7 cm) wide.
Fruits: The fruits are club-shaped and green. When they are ripe, they pop open when touched.
Occurrence: New Guinea impatiens are frequently used in gardens in the tropics and in more moderate climates. As the name indicates, they were originally from New Guinea.
Other names: English "Sunshine Impatiens"; German "Fleißiges Lieschen," "Neuguinea-Impatiens."
Additional information: This species was discovered in 1886, but it was not used in gardens at that time. However, when particularly colorful and resistant species were discovered in 1960, planned breeding started. Today, there are species with different patterned leaves and flowers in all colors from white to violet. Some have two colors. These new species almost pushed the African impatiens out of the market. The African variety was known as "busy Lizzie" or *I. walleriana*. It has very similar flowers, but its growth is looser, and its leaves are alternating.

False Hop

Justicia brandegeana
Family: *Acanthaceae*, Acanthus

Most important features: The flowers are white to pale violet with purple spots on the labium. They grow on drooping spikes that are reddish with a greenish tip.
Growth structure: This is a richly branched herb that is slightly woody at the bottom. It is 12 to 20 inches (30 to 50 cm) high. Occasionally it grows up to 48 inches (120 cm) high.
Leaves: The leaves are set opposite. They are ovate, tapering, and spherical at the bottom. They are 1.6 to 3.6 inches (4 to 9 cm) long and .8 to 2 inches (2 to 5 cm) wide with soft hair.
Flowers: The flowers grow on spikes that are 1.2 to 4.8 inches (3 to 12 cm) long. The spikes are densely populated with ovate bracts that are .4 to .8 inch (1 to 2 cm) long. The corolla is approximately 1.4 inch (3.5 cm) long. It is deeply split into an undivided labrum or one with two lobes and a labium with three lobes.
Fruits: The fruits grow on petioles as ovate capsules that are about .4 inch (1 cm) long. When they are ripe, two flaps open, releasing two to four seeds.
Occurrence: One finds the false hop in gardens. It originated in Mexico.
Other names: English "Mexican Shrimp Plant"; German "Zimmerhopfen," "Garnelenblume"; French "Chevrette"; Spanish "Carpintero," "Cola de Camarón."
Additional information: As with many other members of the Acanthaceae family, the bracts of the false hop are far more attractive than the flowers. In addition, the bracts are present almost all year round. Some species have pure yellow or bright red bracts. In the southern part of the United States, the false hop is used to attract hummingbirds.

Bat Flower

Tacca chantrieri
Family: *Taccaceae*, Chiropterophilous plants

Most important features: The flowers are dark violet to almost black. They grow in one simple umbel that has four large bracts at the bottom. These are almost black. Many long filaments grow from the flowers.
Growth structure: The leaves and inflorescences emerge from one crawling rootstock that is 8 to 28 inches (20 to 70 cm) long.
Leaves: The leaves grow on long petioles. They taper at both tips. The leaves are shiny green with a firm midrib. They are 8 to 22 inches (20 to 55 cm) long and 1.8 to 8.8 inches (4.5 to 22 cm) wide. Often, they are slightly wavy between the lateral veins.
Flowers: The flowers are .4 to .8 inch (1 to 2 cm) long with curled petals and a lower ovary. The flowers grow in simple umbels with five to twenty-five flowers. The two outer bracts are sessile, .8 to 3.6 inches (2 to 9 cm) long, and .4 to 1.6 inch (1 to 4 cm) wide. They are directed toward the top and bottom. The inner ones are wider and ovate to almost heart-shaped. They point sideways on a very short petiole, or they are sessile.
Fruits: The fruits are elliptical, spherical to three-sided, and slightly fleshy. They are .8 to 1.6 inch (2 to 4 cm) long and .4 to .8 inch (1 to 2 cm) thick. The fruits are firm and orange red to purple. Inside are many seeds.
Occurrence: One finds bat flowers in shady places in gardens. Originally, they grew in Southeast Asia.
Other names: English "Black Lily," "Cat's Whiskers," "Devil Flower"; German "Fledermausblume."
Additional information: The bat flower is often planted because of its almost black flowers. These are very odd looking. People in Asia eat the newest leaves. *T. integrifolia*, known as the white bat flower, is very similar, but it has light, erect bracts in the inflorescence. People use the tuber of *T. leontopetaloides* (Tahiti arrowroot) to produce starch, but only after washing out the poisonous, bitter substances.

Blue Ginger

Dichorisandra thyrsiflora
Family: *Commelinaceae*, Dayflower

Most important features: This is an herbaceous plant with firm, bamboo-shaped stems. The leaves have a tubular leaf sheath. The flowers are violet. They have six yellow stamens. The two or three lower ones are longer than the upper ones.
Growth structure: The blue ginger grows erect, up to 7 feet (2 m) high.
Leaves: The leaves are alternating. They accumulate at the ends of the stems. The leaves are tapered at both ends. They are shiny with a firm midrib. The leaves are 6 to 16 inches (15 to 40 cm) long and 2 to 5.2 inches (5 to 13 cm) wide. On the bottom, they are often slightly violet.
Flowers: The flowers grow in groups of three in small panicles at the ends of the stems. The panicles are 2 to 8 inches (5 to 20 cm) long. The sepals are violet on the outside and white on the inside. The petals are violet with white petioles that are .6 inch (15 mm) long.
Fruits: The fruits are elliptical. When they are ripe, they open with three flaps. The seeds have an orange red coat.
Occurrence: The blue ginger is a popular garden plant for semishady areas, especially in Australia and Hawaii. Originally, it was from Brazil.
Other name: German "Blauer Ingwer."
Additional information: Actually, the name of this species is confusing; the blue ginger is not related to ginger. The deep blue violet flower makes it a popular color accent. However, it is very tricky to grow because it needs fertile soil and enough water. It cannot withstand temperatures below 50°F (10°C). On the other hand, it cannot tolerate temperatures above 80°F (27°C) for long periods. The striking *Cochliostema odoratissimum* (see photo at right) is from the same family. It grows as a large, trunkless rosette. In the wild, it often grows on trees. Its flowers are similar to the blue ginger's, but they are about twice as large. They are a lighter violet and grow in dense bunches in the axils of the leaves. They are attached to short petioles.

Boat Orchid

Spathoglottis plicata
Family: *Orchidaceae*, Orchids

Most important features: The flowers grow in erect simple racemes that are 1.5 to 7 feet (0.5 to 2 m) long. The individual flowers are pink to violet red (occasionally white) with five petals. The petals vary in size from elliptical to ovate. A crooked column grows in the upper part of the flower and the lower part of the flower has three lobes.
Growth structure: This is a firm, herbaceous plant with a tuberous bulge at the bottom of each group of leaves.
Leaves: All the leaves start from the ground. At the top, they hang on indistinct petioles. The sheath is often violet, slender, and oblong. The leaves are 24 to 48 inches (60 to 120 cm) long and 2 to 8 inches (5 to 20 cm) wide. They are tapered between the parallel veins that run lengthwise.
Flowers: Between five and twenty-five flowers grow in each simple raceme. The flowers are 1.2 to 2 inches (3 to 5 cm) wide. The three outer petals are slightly smaller than the two inner ones. The lip is dark violet red toward the top, pointing toward the lateral lobes. One is a fleshy yellowish formation with two lobes; the other is a terminal lobe with a petiole.
Fruits: The fruits are green, cylindrical capsules. They are up to 2 inches (5 cm) long and 1.2 inch (3 cm) thick with three light green, longitudinal stripes. When ripe, they open along the stripe. Inside are many tiny seeds.
Occurrence: The boat orchid is native to Southeast Asia. It is still one of the most frequently seen orchids in that area. In Hawaii, it often grows wild in grasslands and along the sides of roads.
Other names: English "Grapette," "Philippine Ground Orchid," "Palm Orchid"; German "Gefaltete Spatelzunge."
Additional information: In the Asian mountains, one can also find other yellow-flowering *Spathoglottis* species growing along the sides of roads.

Flamingo Lily

Anthurium andraeanum
Family: *Araceae*

Most important features: The leaves are large and triangular to heart-shaped. The inflorescences have a bright red, shiny bract at the bottom of a yellowish cob. The bract looks as if it were made of plastic.
Growth structure: The flamingo lily grows as an herbaceous plant without an erect stem. All the leaves have long, firm petioles starting from the ground.
Leaves: The leaves are alternating and tapered. They are 8 to 16 inches (20 to 40 cm) long and 4 to 6.4 inches (10 to 16 cm) wide. Most have seven firm veins that start from the stipula. The lower pair runs into the basal lobes of the leaf and continues to branch there.
Flowers: The bracts are heart-shaped and 3.6 to 4.8 inches (9 to 12 cm) long. They are deeply incised at the bottom, netlike, and often slightly dented between the veins. The spikes are 2.4 to 4 inches (6 to 10 cm) long. They may be erect or drooping. The flowers themselves are tiny, only recognizable as white spots.
Fruits: The fruits are yellow, obovate, berries densely crammed on the cob. The bracts are partly greenish during the flowering season.
Occurrence: The flamingo lily is widely cultivated outdoors and as an indoor plant. The wild specimen grows as an epiphyte (see page 18). It was originally from Colombia.
Other names: English "Tail Flower"; German "Große Flamingoblume"; Spanish "Cresta de Gallo," "Lengua del Diablo."
Additional information: The flamingo lily is one of the most popular ornamental herbaceous plants found in shady localities. When cultivated, the specimens are often larger with pink or white bracts. *A. scherzerianum* is even more popular as an indoor plant. It is slightly more robust and differs from the flamingo lily because of its tapering, slender leaves, its smooth, ovate, hardly incised bracts, and its cob, which may be bent or wound.

Philippine Wax Flower

Etlingera elatior
Family: *Zingiberaceae*, Ginger

Most important features: The flowers grow in waxlike bracts with a light margin. The lower ones are clearly larger than the upper ones. The bracts range in color from pink to bright red and have a white or light margin. The plants are often leafless. They grow from 3 to 7 feet (1 to 2 m) high.
Growth structure: The Philippine wax flower is a very large herbaceous plant with firm stalks. It grows 7 to 20 feet (2 to 6 m) high.
Leaves: The leaves are deeply divided with a long sheath and a short petiole. They are 12 to 35 inches (30 to 90 cm) long and 4.8 to 8 inches (12 to 20 cm) wide. The leaves are tapering with a firm midrib and many delicate lateral veins.
Flowers: Often, only the lip of the flower is visible. The flowers may be red with yellow or white limbs. The flowers are almost tubular and open at the upper side. Only one level of the flower head is open at a time. The lower bracts have no flowers. The flowers themselves grow up to 4.8 inches (12 cm) long. The flowers are larger at the bottom than they are at the top.
Fruits: The fruits are spherical, slightly fleshy, and green to reddish. They are approximately .8 to 1 inch (2 to 2.5 cm) long and about 4 inches (10 cm) wide.
Occurrence: One finds the Philippine wax flower as an ornamental herbaceous plant in all tropical areas. Originally, it was from the Indo-Malaysian region.
Other names: English "Torch Ginger"; German "Fackelingwer," "Kaiserzepter"; French "Rose de Porcelaine"; Spanish "Boca de Dragón."
Additional information: Scientists are still not in agreement about the scientific name of the Philippine wax flower. In addition to *Etlingera elatior*, it is also called *Nicolaia elatior* or *Phaeomeria magnifica*. Young inflorescences are eaten uncooked in salads in Southeast Asia. Other ginger plants have similar, but smaller and less striking inflorescences. Goldenseal (*Curcuma longa*) is used in making curry. Its bracts and flowers are white with green or white with yellow.

Giant Spiral Ginger

Tapeinochilos ananassae
Family: *Zingiberaceae*, Ginger

Most important features: The stalks are shaped like bamboo. They are light brown to reddish with green knots. The foliage is often curled and spiral-like. The flowers are yellow in bright red, cone-shaped spikes, which often grow at leafless shoots in the shadow of the leaves.
Growth structure: The plant grows erect with a firm rootstock. It grows 4.8 to 13 feet (1.5 to 4 m) high.
Leaves: The leaves are alternating, almost sessile, and ovate to tapering. They grow 3.2 to 10 inches (8 to 25 cm) long and .6 to 2.8 inches (1.5 to 7 cm) wide with a reddish brown leaf sheath, clear midrib, and delicate lateral veins.
Flowers: The flowers are asymmetrical with one stamen. They are shorter than the hard, tapering, waxlike bracts. Frequently, only a few are open at the same time. The spikes are 2.8 to 8 inches (7 to 20 cm) long and 2.8 to 4 inches (7 to 10 cm) thick on stems that are 8 to 80 inches (20 to 200 cm) long. The stems emerge directly from the soil.
Fruits: These are hidden between the bracts of the inflorescence. From the outside, only the blackened remainders of the petals are visible.
Occurrence: Giant spiral gingers grow in semi-shade in parks and gardens. Originally, they spread from Indonesia to northern Australia.
Other names: English "Indonesian Wax Ginger," "Pineapple Ginger"; German "Ananas-Ingwer."
Additional information: The giant spiral ginger owes its name to its relationship with the real ginger and to the fact that its inflorescences faintly resemble the pineapple. However, it is not related to the pineapple. It is more closely related to the crape ginger (see page 238), which also has spiral, curled shoots. Several other species have spikes with colored bracts, yet none appears as artificial as the giant spiral ginger, which looks as if it were plastic.

Indian Shot

Canna indica
Family: *Cannaceae*, Canna

Most important features: The flowers are asymmetrical and strikingly colored, often in red and yellow. The ovary is inferior with warts and thorns.
Growth structure: The plant grows erect with a crawling, knotty rootstock. The Indian shot reaches a height of 1.5 to 7 feet (0.5 to 2 m).
Leaves: The leaves are alternating with long leaf sheaths that are inserted into each other. The leaf blade is ovate to tapering with many parallel lateral veins. The leaves are 6 to 24 inches (15 to 60 cm) long and 3.2 to 10 inches (8 to 25 cm) wide. Occasionally, they are red or purple brown.
Flowers: The flowers grow in erect, dense panicles at the ends of the stems. They are more slender in the axil. The flowers are 2 to 3 inches (5 to 7.5 cm) long. Two or three of the petals are larger than the others. The single stamen is flower-shaped and partly adheres to the pistil. The flowers grow in bracts that are 6 inches (15 cm) long.
Fruits: The fruits are spherical with soft thorns. They are .8 to 1.6 inch (2 to 4 cm) long and .6 to 1.2 inch (1.5 to 3 cm) thick. When they are ripe, three flaps open, revealing many black seeds that look like shotgun pellets.
Occurrence: This is one of the most popular ornamental plants. Despite its name, *indica*, it was originally from the Western Hemisphere.
Other names: German "Blumenrohr"; French "Balisier"; Spanish "Bandera Española," "Platanillo," "Yuquilla."
Additional information: The Indian shot is one of the few plants with entirely asymmetrical flowers. Often they are red with yellow spots or the other way around. However, some have wine red, orange, pink, or cream white flowers (photo at right is a wild specimen). The most striking petals are not part of the corolla; they are transformed stamens. The narrow organs outside are petals. The Indian shot cannot withstand frost, but it can be cultivated in moderate climates if one digs up the rootstocks in the fall and keeps them in a frost-free place.

Red Ginger

Alpinia purpurata
Family: *Zingiberaceae*, Ginger

Most important features: The leaves are often arranged in two rows with one leaf sheath. The sheath has an erect ligule at the upper tip. The inflorescences have striking red bracts and comparatively insignificant white flowers.
Growth structure: This is a firm herbaceous plant. It usually grows from 1.5 to 8.5 feet (1.5 to 2.5 m) high, but it may reach 16 feet (5 m).
Leaves: The leaves are tapering, bare, and almost sessile with a short petiole. They are 12 to 32 inches (30 to 80 cm) long and 2 to 8 inches (5 to 20 cm) wide with a distinct midrib and delicate, almost straight lateral veins that branch out at an acute angle. The ligule is .3 to .8 inch (5 to 20 mm) long. It is hairy and often has two unequal lobes.
Flowers: The flowers grow in spikes that are erect or drooping at the tip. The spikes are 6 to 15 inches (15 to 35 cm) long and grow at the end of the shoots. The bracts are 1 to 2.4 inches (2.5 to 6 cm) long. One to five flowers grow in the axils. The flowers are about .8 inch (2 cm) long.
Fruits: The fruits are almost spherical with three flaps. They are .8 to 1.2 inch (2 to 3 cm) long and are seldom seen.
Occurrence: The red ginger is popular as an ornamental plant in all tropical areas. It also often grows wild. Scientists believe it probably spread from Indonesia to the New Hebrides.
Other names: German "Scharlachrote Alpinie"; French "Lavande Rouge"; Spanish "Gengibre Rojo."
Additional information: The red ginger is admired for its striking inflorescences. The bracts of the inflorescences remain almost all year round, even though the flowers live for only a short time. Besides the red specimen, a specimen with pink bracts, pink ginger, is also cultivated. Thus, "ginger" indicates the close relationship to the ginger (*Zingiber officinale*), which is much smaller and has short, dense spikes with green and yellow bracts and yellow flowers with a violet lip.

Beach Morning Glory

Ipomoea pes-caprae
Family: *Convolvulaceae*, Bindweed

Most important features: This is a crawling plant. The ends of the leaves are deeply incised and wide. Rarely are they straight. The flowers are large and pink to red violet with a darker throat. The flowers are only open in the morning.
Growth structure: The beach morning glory grows with herbaceous shoots that are up to 100 feet (30 m) long. It crawls across the sand.
Leaves: The leaves are alternating. They are rectangular and tapering to elliptical. The leaves are 1.2 to 4.8 inches (3 to 12 cm) long and slightly fleshy.
Flowers: These grow individually (occasionally in small groups). The sepals are .2 to .6 inch (0.5 to 1.5 cm) long. The outer ones are smaller than the inner ones. The corolla is funnel-shaped and 1.2 to 2.8 inches (3 to 7 cm) long.
Fruits: The fruits are ovate to flattened and spherical. They are .4 to .8 inch (1 to 2 cm) long with two chambers. When ripe, four flaps open, revealing four large, black, hairy seeds.
Occurrence: One finds the beach morning glory on almost all tropical coasts. Occasionally, one also sees them by the side of large lakes.
Other names: English "Goat's Foot Vine," "Sea Morning Glory"; German "Ziegenfußwinde," "Strandwinde"; French "Patate Marron"; Portuguese "Batata de Mar"; Spanish "Bejuco de Playa."
Additional information: The beach morning glory owes its name to the fact that the flowers unfold before noon and wilt in the afternoon. The plant has spread to almost every tropical area without cultivation; its seeds are capable of floating. The new roots at the knots of the shoot hold the sand together, making beaches firmer. The beach morning glory is similar to many other *Ipomoea* species (see pages 194, 200, 206, and 270). In some areas, the juice of its shoot is used to treat poisonous fish spines.

Dancing Lady Ginger

Globba winitii
Family: *Zingiberaceae*, Ginger

Most important features: The inflorescences have red violet bracts at the main axil and bizarre, hook-shaped, yellow flowers on the lateral branches.
Growth structure: This an herbaceous plant that grows 12 to 35 inches (30 to 90 cm) high.
Leaves: The leaves are alternating, tapering, and often deeply divided with a leaf sheath. The leaves are up to 4 inches (10 cm) long. At the bottom, they are spherical to heart-shaped. The veins grow in very acute angles branching out from the midrib.
Flowers: The flowers grow in drooping, simple racemelike panicles at the end of the stamen. They are 2 to 8 inches (5 to 20 cm) long . The bracts grow up to 1.2 inch (3 cm) long. Often, only one flower unfolds per branch. The calyx is almost horizontal with three tiny cusps. The corolla tube is thin above the calyx. The corolla cusps contain the flat, sterile, and curled stamens. The only fertile stamen is much longer than the corolla. It is shaped like a bent arch with four cusps at the end.
Fruits: The fruits are small, spherical capsules. The seeds have a fringed coat.
Occurrence: One finds this popular ornamental plant in very shady locations. It originated in Thailand.
Other names: English "Mauve Dancing Girl"; German "Globba."
Additional information: Many people believe the dancing lady ginger is an orchid because of its bizarre flowers; however, it is more closely related to the ginger. The other thirty-five *Globba* species have less striking bracts. Their inflorescences are partly erect, and they have orange, red, or violet flowers. Most of them can grow brood buds instead of flowers in the axils of the lower bracts, reproducing themselves without pollination. The brood buds are often mistaken for fruits because they are a striking red, black, or red black color.

Fireweed

Celosia argentea
Family: *Amaranthaceae*, Amaranth

Most important features: The stem is firmly squared. The leaves are narrow and tapering at both ends. The flowers grow in dense spikes at the ends of the stems. The flowers are paperlike and often pink. After flowering, the color rapidly fades.
Growth structure: Fireweed grows erect as an annual herb. It reaches 1.5 to 5 feet (0.5 to 1.5 m).
Leaves: The leaves are alternating. They are 1.6 to 7.2 inches (4 to 18 cm) long and .2 to 2.6 inches (0.5 to 6.5 cm) wide. They often have a reddish midrib. Sometimes, they are streaked with violet.
Flowers: The flowers grow on erect spikes that are .8 to 8.8 inches (2 to 22 cm) long. The spikes are slender and branched. The flowers are .3 inch (8 mm) long with a simple perianth. The filaments are connected at the bottom.
Fruits: These are spherical and enclosed in the dry perianth. They are .2 inch (5 mm) long with black seeds.
Occurrence: One can find fireweed anywhere in the tropics. Although they often grow as weeds, they are also used as ornamental plants. They probably originated in Africa.
Other names: German "Celosie"; English "Quailgrass," "Red Fox"; Spanish "Amor Seco," "Boria," "Moño."
Additional information: One frequently finds fireweed plants growing wild on fields that have recently been destroyed by fire (hence, fireweed). The young shoots can be eaten as vegetables. The roots are said to have medical properties. The cultivated specimens one finds in a garden are different than the wild specimens, whose inflorescences are often strongly branched or flat, widened, and wavy. Cockscomb (*var. cristata*, see photo at right) is an example of a wild species. They exist in all colors, from white to yellow and orange, up to dark wine red.

Congo Cockatoo

Impatiens niamniamensis
Family: *Balsaminaceae*, Touch-me-not

Most important features: The stem is glassy and fleshy. The flowers have short petioles and are often green in front. They are yellow and red toward the back. Occasionally, they are white in front or entirely red. At the bottom, they have a spur that is bent toward the front.
Growth structure: This plant grows erect up to 12 to 35 inches (30 to 90 cm), but it may reach 7 feet (2 m).
Leaves: The leaves are alternating, tapering at both ends, and ovate to elliptical with long petioles. The leaves are 2 to 8.8 inches (5 to 22 cm) long and 1.2 to 3.6 inches (3 to 9 cm) wide with a scalloped margin.
Flowers: The flowers grow individually or in groups of up to eight. They are .6 to 1.2 inch (1.5 to 3 cm) long with two tiny sepals on the side. One is sac-shaped and widened. The lower sepal has a spur that creates the larger part of the flower. The three light green corolla lobes are about .4 inch (1 cm) long. One is shaped like a bonnet and points in an upward direction. The other two point downward. The stamen tube is short and yellow. It is bent toward the ground.
Fruits: The fruits are spindle-shaped and green. They are about .6 inch (1.5 cm) long and .2 inch (0.5 cm) thick. The ripe fruits burst open when they are touched.
Occurrence: Congo cockatoo grows in shady, humid places. Originally, it was from tropical Africa.
Other name: German "Kongo-Springkraut."
Additional information: With its bizarrely shaped, often green, yellow, and red flowers, the Congo cockatoo attracts birds, which suck nectar out of the spur, thus pollinating the flowers. Unlike hummingbirds, the flower birds that pollinate the Congo cockatoo cannot hover in the air. Consequently, they have to land on the flower in order to drink. Therefore, the flowers are located close to the stem.

Kahili Ginger

Hedychium gardnerianum
Family: *Zingiberaceae*, Ginger

Most important features: The leaves grow in two vertical rows with one erect appendage at the end of the leaf sheath. The appendage is .6 to 1.6 inch (1.5 to 4 cm) long. The flowers grow in regular, dense, erect spikes. They are golden yellow and have a wonderful aroma. Each flower has one red stamen.
Growth structure: This large herbaceous plant with a firm rootstock grows 3 to 7 feet (1 to 2 m) high.
Leaves: The leaves are sessile, oblong, tapering, and hairy on the bottom. They are 8 to 18 inches (20 to 45 cm) long and 1.6 to 6 inches (4 to 15 cm) wide.
Flowers: The flowers grow on spikes at the ends of the stems. The spikes are 10 to 18 inches (25 to 45 cm) long. The flowers grow individually or in pairs in the axils. The flowers are 1.2 to 2.8 inches (3 to 7 cm) long with narrow bracts. The corolla tube is 1.4 to 2.4 inches (3.5 to 6 cm) long. The tube is thin at the end. It has three very narrow cusps and three wide cusps that are 1 to 1.6 inch (2.5 to 4 cm) long. The largest of the cusps is .6 to .8 inch (1.5 to 2 cm) wide and undivided up to two lobes. The single stamen is longer, up to 1.8 inch (4.5 cm) long.
Fruits: The fruits are tapering to elliptical and orange inside. They grow up to 1.2 inch (3 cm) long. When they are ripe, they open with three flaps. The coat of the seeds is scarlet red.
Occurrence: The Kahili ginger grows as a garden plant in many tropical and subtropical areas. It originated in the eastern region of the Himalayas.
Other names: German "Gelbe Hedychie," "Zier-Ingwer"; Portuguese "Roca de Velhac."
Additional information: The Kahili ginger grows wild in altitudes up to 8,200 feet (2,500 m). It is quite robust and can grow outside in moderate latitudes if the rootstock is stored in a frost-free place during the winter months. It was introduced worldwide as an ornamental plant; however, it is now considered a weed in areas such as the Azores, Hawaii, and New Zealand. The wide leaves in its flowers are actually transformed stamens; the slender cusps are the real petals. In the tropics, one also see the white gingerlily (*H. coronarium*, see photo at right).

Crab's Claws

Heliconia rostrata
Family: *Musaceae*, Banana

Most important features: This is a gigantic herbaceous plant with hanging inflorescences. The large, predominantly red bracts are yellow at the tip and green at the margin.
Growth structure: The plant grows erect, 7 to 16 feet (2 to 5 m) high with a firm rootstock.
Leaves: The leaves have a long petiole and a firm midrib. They are 1.5 to 7 feet (0.5 to 2 m) long and 6 to 16 inches (15 to 40 cm) wide. The leaves are spherical at the bottom. Between the steeply branching lateral veins, the leaves are often pinnately torn.
Flowers: The inflorescences are 12 to 40 inches (30 to 100 cm) long and hang down from 3 to 10 feet (1 to 3 m). The midrib is twined. The twelve to thirty-five bracts in the inflorescence are very firm and 2.4 to 6 inches (6 to 15 cm) long and .8 to 3.2 inches (2 to 8 cm) wide. They are folded and boat-shaped. The bracts are .4 to 1.2 inch (1 to 3 cm) long and grow on petioles. The flowers are tubular, slightly crooked, and 1.4 to 2.2 inches (3.5 to 5.5 cm) long. They are yellow, often pale at the bottom, and greenish at the tip.
Fruits: The fruits are spherical to triangular, about .4 inch (1 cm) long, and slightly rounded at the end. They start out yellow, but turn violet when they are ripe. Inside are three pits.
Occurrence: These are cultivated everywhere in the tropics, but they seldom grow wild. Originally, they were from South America.
Other names: German "Geschnäbelte Heliconie"; English "Hanging Lobster Claws"; French " Bec de Perroquet."
Additional information: Although each bract encompasses many flowers, usually only two are flowering at the same time. Birds pollinate the flowers when they land on the bracts in order to suck the nectar. The plant only thrives in lowlands because it cannot withstand temperatures below 60°F (15°C). Crab's claws appears to be very similar to a banana tree when it is without flowers; and like bananas, it develops a false trunk out of leaf sheaths that are inserted into each other.

Parakeet Flower

Heliconia psittacorum
Family: *Musaceae*, Banana

Most important features: The parakeet flower is an herbaceous plant with one stem that grows out of leaf sheaths that are inserted into each other. The inflorescences are erect with two to six bright orange bracts. These can also be red at the tip. The axil of each bract has several tubular flowers that are slightly crooked.
Growth structure: The plant grows erect with a crawling rootstock. It is often 3 feet (1 m) high and may grow up to 10 feet (3 m) high.
Leaves: The leaves grow in two vertical rows. The lower ones have long petioles; the upper ones are almost sessile. The leaves are tapering and leathery. They are 6 to 24 inches (15 to 60 cm) long and 2.4 to 4.8 inches (6 to 12 cm) wide with a firm midrib. They are often spherical on the bottom.
Flowers: The bracts are 1.2 to 6 inches (3 to 15 cm) long and boat-shaped. The flowers grow up to 1 to 2 inches (2.5 to 5 cm) long on petioles that are .8 inch (2 cm) long. The flowers are slender and orange, yellow, or greenish. The tip is green to almost black.
Fruits: The fruits are spherical to cone-shaped and slightly squared. They are up to .4 inch (1 cm) long. They are rounded at the end. At first, they are yellowish, but they turn dark blue when they are ripe. Each fruit has three pits.
Occurrence: One finds the parakeet flower anywhere in the tropics. Originally, it was from Guyana and Brazil.
Other names: English "Parrot Flower," "Parrot's Plantain"; German "Papageien-Heliconie"; French "Petit Balisier."
Additional information: The parakeet flower is probably the most popular of the more than two hundred *Heliconia* species. Many of these are used as ornamental plants. In nature, one finds them in humid savannas, yet they are also often found along roadsides. The parakeet flower is cultivated and crossbred with closely related species in a variety of colors from pale yellow to pink up to deep red, often without the characteristic dark flower tip. The entirely orange specimen, golden torch, and specimens with red bracts are especially popular.

Bird of Paradise

Strelitzia reginae
Family: *Musaceae*, Banana

Most important features: The leaves are shaped like bananas, but they are bluish or gray green. The inflorescences have one boat-shaped bract. One to three orange and violet flowers stand out at the top of the bract.
Growth structure: This is a firm herbaceous plant that grows up to 7 feet (2 m) high.
Leaves: The leaves grow in two vertical rows on petioles that are up to .4 inch (1 cm) long. The leaf blade is 8 to 28 inches (20 to 70 cm) long and 2.4 to 6 inches (6 to 15 cm) wide. It is tapering, stiff, and leathery with a very strong midrib and many delicate lateral veins.
Flowers: These bloom one after the other. Each has a maximum of three petals that are 3.2 to 4.8 inches (8 to 12 cm) long. The petals are delicate. They are orange and violet. Two of the three violet ones clasp the pistil and the five stamens. The third violet petal is much smaller. The bract is 4.8 to 7.2 inches (12 to 18 cm) long. Often, it is almost at a right angle to the petiole. The bract is gray green. Usually, it is reddish toward the margin.
Fruits: The fruits are leathery and have three flaps with many seeds, each with an orange appendage.
Occurrence: The bird of paradise is frequently cultivated in all tropical and subtropical areas. It originated in South Africa.
Other names: English "Crane Flower"; German "Paradiesvogelblume"; French "Oiseau de Paradis"; Spanish "Flor Ave del Paraíso."
Additional information: Although the folk name refers to the bright colors, not to a bird, the plant is actually pollinated by birds. They land on the long violet petals, pushing them apart so that stamens and pistil touch the birds' feet or breast feathers. The closely related *S. nicolai* is pollinated the same way. Its flowers are white and can grow up to 33 feet (10 m) high. Sir Joseph Banks used the name, *Strelitzia*, to honor the wife of King George III of England, Charlotte von Mecklenburg-Strelitz.

Devil's Fig

Argemone mexicana
Family: *Papaveraceae*, Poppy

Most important features: All of the green parts of the plant have thorns and a yellow, milky sap. The flowers are yellow with many stamens and a purple stylar head.
Growth structure: The plant grows erect with few branches. It reaches a height of 6 to 52 inches (15 to 130 cm).
Leaves: The leaves are alternating with deep pinnate lobes. The lobes terminate in pointed thorns. The edges are often rolled backward. The upper leaves are sessile and clasp the stem. The leaves are up to 8 inches (20 cm) long and 2.8 inches (7 cm) wide. They are bluish green with white veins.
Flowers: The flowers grow individually at the ends of the stems. They are 1.6 to 3.2 inches (4 to 8 cm) long. Each flower has three sepals, which are shed early, and six petals. These are wrinkly in the bud, but then spread out in a bowl shape.
Fruits: The fruits are spherical to tapering and 1 to 1.4 inches (2.5 to 3.5 cm) long. They have three to six flaps and many very small seeds.
Occurrence: Devil's fig grows as a weed in all tropical and subtropical areas. It originated in Central America.
Other names: English "Mexican Poppy," "Prickly Poppy"; German "Stachelmohn"; Portuguese "Cardo Santo"; Spanish "Cardo Amarillo," "Chicalote."
Additional information: The Devil's fig is remotely related to the poppy (*Papaver*). It was introduced into many areas as an ornamental plant; however, it turned out to be weed. Because it is thorny and poisonous, few animals eat it. In addition, it spreads very rapidly because the seeds are ripe within a few months of germination. It prefers open and dry locations and tends to drive out the weaker, native vegetation. In earlier times, people pressed the oil out of the seeds and used it for medical purposes.

Bahama Buttercup

Turnera ulmifolia
Family: *Turneraceae*, Damiana

Most important features: The leaves are alternating and nettle-like. The flowers are about 2 inches (5 cm) long with five yellow petals and five stamens. The three pistils end with fringed stigmas.
Growth structure: This is an herbaceous plant to subshrub. It grows 1.5 to 5 feet (0.5 to 1.5 m) high.
Leaves: The leaves are elliptical to very narrow. They are hairy, especially on the bottom, and coarsely serrated on the margin. The leaves are 3.2 to 6 inches (8 to 15 cm) long and .6 to 2 inches (1.5 to 5 cm) wide.
Flowers: The flowers grow individually or in small groups in the uppermost axils of the leaves. The sepals are about .4 inch (1 cm) long. The five petals are .8 to 1 inch (2 to 2.5 cm) long. They are shaped like a wide spatula. At the end of the petal, there is often a small cusp, or the petal will be slightly scalloped. The flowers are pale to bright yellow. Often the bottom has a dark, almost black spot.
Fruits: The fruits are green, elliptical to cone-shaped capsules. They are approximately .2 inch (5 mm) long.
Occurrence: One finds the Bahama buttercup as an ornamental plant or as a weed in all tropical areas. Originally, it grew in the Western Hemisphere.
Other names: English "Sage Rose," "Large-Leaf Damiana," "West Indian Holly," "Yellow Alder"; German "Großblättrige Damiana"; French "La Coquette"; Portuguese "Albina," Chanana"; Spanish "Marilope."
Additional information: The Bahama buttercup encompasses many slightly different specimens. Sometimes, these are considered separate species. Some people believe that the tea made from its leaves is a strong diuretic and can relieve spasms. The real damiana (*T. diffusa*) is thought to have an even stronger effect. Because it was supposed to have such strong healing properties, the plant was named in honor of St. Damian, the patron of doctors, pharmacists, and barbers.

Nun's Orchid

Phaius tankervilliae
Family: *Orchidaceae*, Orchids

Most important features: The large flowers grow in one firm, erect simple raceme. The raceme is 3 to 5 feet (1 to 1.5 m) long with five narrow petals that are white on the outside and light brown on the inside. The petals are curled, and part of the lip of the petal is purple red.
Growth structure: Nun's orchid is a firm herbaceous plant with tuberous bulges at the bottom of one group of leaves.
Leaves: All of the leaves grow from the bottom. They are erect with an indistinct, segmented petiole. The leaves are narrow, oblong, and tapering. They are 24 to 40 inches (60 to 100 cm) long. Between the veins, the leaves are grooved lengthwise and folded.
Flowers: These point sideways with firm petals that stand off diagonally. They are 3.2 to 4.8 inches (8 to 12 cm) long. The lip is wrapped around a midrib. At the bottom, it is wrapped with a hook-shaped spur; at the end, with a slightly lengthened midlobe. It is yellowish at the bottom, white in the middle, and reddish at the end; or it is entirely red and lighter at the end.
Fruits: The fruits hang down. They are grooved capsules with many tiny seeds. One seldom sees the fruits.
Occurrence: One finds nun's orchids in humid grasslands and in gardens. Sometimes, they grow wild. They have spread naturally from India to the southern part of China to New Guinea and the northern part of Australia.
Other names: English "Veiled Nun Orchid"; German "Nonnenorchidee."
Additional information: The nun's orchid belongs to the largest group of orchids that live on the ground. Most of the others, more than 18,000 species of this family, grow on trees (see pages 246 to 250). Although it originally grew in relatively good soil at altitudes of 2,000 to 4,200 feet (610 to 1,300 m), it has spread to many places in the lowlands.

American Aloe

Agave americana
Family: *Agavaceae*, Agave

Most important features: The leaves are oblong, thick, and fleshy with a thorny pointed end and thorns on the margin. Between the ends, it is slightly lobed. The inflorescence is gigantic and rich in flowers.
Growth structure: This is a large, firm rosette without a trunk. The young leaves are erect; often the older ones are curled upward and then downward.
Leaves: The leaves are waxy and bluish or gray green. Sometimes, they have light longitudinal stripes. The leaves are 3.2 to 8.8 feet (1 to 2.5 m) long, 4.8 to 12 inches (12 to 30 cm) wide, and 1.6 to 3.2 inches (4 to 8 cm) thick. They are widest in the middle. The thorns have a wide base and are often bent upward.
Flowers: The flowers grow on the lateral branches of a single inflorescence that grows 16 to 46 feet (5 to 14 m) high. The flowers are 3.2 to 4 inches (8 to 10 cm) long and grow in groups of six. They are yellowish green. The inferior ovary is relatively large.
Fruits: When ripe, the fruits, elliptical to tapering, open with three flaps, revealing many seeds.
Occurrence: American aloe is cultivated in all tropical and subtropical areas and often grows wild in these areas. Originally, it came from Central America.
Other names: English "Century Plant"; German "Amerikanische Agave"; Spanish "Maguey," "Pita Común."
Additional information: The water-saving leaves of the aloe are a clear indication that the plant originated in dry areas. However, it is also cultivated in the humid tropics. After eight to twenty years, the rosette flowers only once. It dies after its fruits ripen. However, before it dies, the plant creates offshoots that start growing. The juice of several aloe species is rich in sugar and can be fermented. The sisal hemp (*A. sisalana*) supplies woody fibers up to 3 feet (1 m) long that are used to produce rope.

Pink Porcelain Lily

Alpinia zerumbet
Family: *Zingiberaceae*, Ginger

Most important features: The leaves are arranged in two vertical rows with one leaf sheath carrying an erect ligule at the upper end. The inflorescences are drooping. The bract whorls are white with a pink tip. Only the lip is larger. The flower is yellow with red stripes and a red throat.
Growth structure: This is a firm, herbaceous plant with a strong rootstock. It grows 7 to 13 feet (2 to 4 m) high.
Leaves: The leaves are tapering and almost sessile. They are 12 to 28 inches (30 to 70 cm) long and 2.4 to 6 inches (6 to 15 cm) wide. Young leaves are hairy on the margin. The leaves have a distinct midrib. The lateral veins are delicate and almost straight. The ligule is .4 to 1.2 inch (1 to 3 cm) long and hairy.
Flowers: The flowers grow in simple drooping racemes at the ends of the shoots without striking bracts. The racemes are 4 to 12 inches (10 to 30 cm) long. The perianth is .8 to 1.6 inch (2 to 4 cm) long, and the lip is up to 2 inches (5 cm) long.
Fruits: The fruits are spherical and orange red with a berry shape. They are about .8 inch (2 cm) long. When they ripen, three flaps open.
Occurrence: This is a popular ornamental plant in all tropical areas. The exact place of origin is uncertain, but the most likely area is Southeast Asia.
Other names: English "Shell Ginger"; German "Nickende Alpinie"; French "Lavande Blanche," "Grand Dégonflé"; Portuguese "Colônia"; Spanish "Lírio de Colón."
Additional information: The pink porcelain lily owes its name to the bract whorls, which are as transparent as porcelain. The three outer whorls are shell-like and overgrown. As with all ginger plants, the lip is created by two transformed stamens. There are more than two hundred *Alpinia* species. These include red ginger (see page 224) and the greater galangal and lesser galangal (*A. galanga* and *A. officinarum*), both of which are used as spices in Asia.

Crape Ginger

Costus speciosus
Family: *Zingiberaceae*, Ginger

Most important features: The flowers grow in dense, erect spikes at the ends of the stems. Often, only a few open at the same time. Each flower has only one very large, coiled petal. It is white with a yellow center and slightly fringed on the margin.
Growth structure: Crape ginger is a large herbaceous plant with a firm rootstock. It grows up to 10 feet (3 m) high. The stem may be drooping or grow like a spiral. When it spirals, all the leaves are turned outward.
Leaves: The leaves are alternating, tapering, shiny, and hairy on the bottom. They are 5.6 to 18 inches (14 to 45 cm) long and 1.6 to 4 inches (4 to 10 cm) wide with a short petiole and lateral veins branching from the midrib at an acute angle.
Flowers: The flowers grow on spikes that are 1 to 4.4 inches (2.5 to 11 cm) long and 1 to 3.2 inches (2.5 to 8 cm) thick. The bracts and sepals are reddish and tapering. The petals are white and 1.2 to 2 inches (3 to 5 cm) long. They are small compared with the funnel-shaped lip, which is 2 to 3.6 inches (5 to 9 cm) long.
Fruits: The fruits are bright red, elliptical, and hairy. They are .4 to 1.2 inch (1 to 3 cm) long and crowned by the calyx. When ripe, three clefts open.
Occurrence: The crape ginger is native to tropical Asia, where it is very widely spread
Other names: English "Malay Ginger," "Spiral Ginger," "White Costus"; German "Kostwurz"; Spanish "Caña Mejicana."
Additional information: The crape ginger plays an important role in Asian folk medicine. Although wild specimens usually grow erect, in cultivation, one often finds specimens that are strongly spiral-like. Typically, the leaves of these specimens have yellow stripes. Of the approximately forty *Costus* species, some have yellow, orange, or red flowers and are kept as ornamental plants.

Peace Lily

Spathiphyllum floribundum
Family: *Araceae*

Most important features: The leaves have a leaf sheath running down the petiole as a limb. They also have a leaf joint shortly below the blade. The inflorescence is white or partly greenish. It spreads out of a bract at the bottom of a yellowish cob.
Growth structure: Peace lily is an herb with a crawling shoot. It grows 8 to 18 inches (20 to 45 cm) high.
Leaves: Almost all of the leaves start from the ground. They are tapering and ovate up to very narrow. The leaves grow on petioles that are 4 to 8 inches (10 to 20 cm) long. The leaves are 4.8 to 10 inches (12 to 25 cm) long and 1.2 to 2.8 inches (3 to 7 cm) wide with a firm midrib and slightly wavy lateral veins.
Flowers: The flowers are tiny and densely packed on a cob that is .8 to 3.2 inches (2 to 8 cm) long and .2 to .3 inch (5 to 8 mm) thick. The cob grows on a petiole that is 8 to 15 inches (20 to 35 cm) long. The bract is 2.4 to 4 inches (6 to 10 cm) long, .8 to 1.2 inch (2 to 3 cm) wide, and tapering. The midrib and pointed end are often greenish.
Fruits: The fruits are small, ovate, and yellowish to orange. They are densely packed on the cob.
Occurrence: The peace lily is found in gardens. It originated in Panama and Colombia.
Other names: English "Snow Flower," "Spathe Flower," "White Flag"; German "Blattfahne," "Scheidenblatt," "Einblatt."
Additional information: The peace lily belongs to one of the most popular of all ornamental plants found in moderate latitudes. In the tropics, it is frequently planted in semishady areas of the garden. Because of its high content of oxalic acid, it is poisonous. Several others of the thirty-six *Spathiphyllum* species are cultivated. Almost all of them originate in the Western Hemisphere and differ only slightly from each other. One often sees *Sp. cannaefolium* with its leathery leaves. *Sp. wallisii* can grow taller than 3 feet (1 m) high, as can the Asian *Sp. Commutatum.*

Calla Lily

Zantedeschia aethiopica
Family: *Araceae*

Most important features: The leaves are heart-shaped to triangular. The inflorescence has one white bract that is greenish at the bottom. The bract is coiled in a funnel shape around a light yellow cob.
Growth structure: The calla lily is an herbaceous plant with a firm rootstock. It grows 24 to 60 inches (60 to 150 cm) high.
Leaves: All the leaves start from the ground. They grow on slightly fleshy petioles that are up to 24 inches (60 cm) long. The leaf blade is 6 to 20 inches (15 to 50 cm) long and 4 to 10 inches (10 to 25 cm) wide. The leaf has a firm midrib and many parallel lateral veins. These are not equally strong.
Flowers: The bracts are 4.8 to 10 inches (12 to 25 cm) long with a delicate, drooping tip. The cob is 2.4 to 3.2 inches (6 to 8 cm) long and erect. The flowers themselves are tiny and hard to see.
Fruits: The fruits are spherical berries enclosed in the remainders of the bracts.
Occurrence: Calla lilies are cultivated in many tropical and subtropical areas. In addition, they may grow wild in marshy areas. Originally, they were from South Africa.
Other names: English "Pig Lily," "Trumpet Lily," "White Arum Lily"; German "Zantedeschie," "Zimmerkalla"; French "Pied de Veau"; Portuguese "Copo de Leite"; Spanish "Cala Blanca."
Additional information: The plant is neither a calla, to which it is at best distantly related, nor a lily. It thrives in any area where there is enough humidity and where there is not a frost that is strong enough to reach the rootstock hidden in the marsh. This is a poisonous plant. It grows rapidly out of each fragment of its rootstock. In Australia and New Zealand, it is considered a very undesirable weed.

Poison Lily

Crinum asiaticum
Family: *Amaryllidaceae*, Amaryllis

Most important features: Poison lily is a firm herbaceous plant with long leaves. The large, white flowers grow in groups of six in simple umbels with firm, slightly flattened petioles on the side.
Growth structure: The plant grows up to 7 feet (2 m) high with an oblong bulb that is often up to 32 inches (75 cm) high and a stemlike bulbous neck.
Leaves: The leaves are lightly fleshy and taper at the end. They are 12 to 60 inches (30 to 150 cm) long and 1.6 to 8 inches (4 to 20 cm) wide with parallel veins.
Flowers: Between ten and forty flowers grow on each simple umbel. The petiole of the umbel is 18 to 32 inches (45 to 80 cm) long. The corolla tube is 2.4 to 5.2 inches (6 to 13 cm) long. The corolla lobes are 2.4 to 4.8 inches (6 to 12 cm) long and .2 to .6 inch (5 to 15 mm) wide. They are spread out or curled downward and streaked with white or red. The stamens are 2 to 4 inches (5 to 10 cm) long and violet at the end. The ovary is inferior.
Fruits: The fruits are spherical and green. They are 1.8 to 2.6 inches (4.5 to 6.5 cm) long with two to four large, green seeds.
Occurrence: Poison lily is cultivated in all tropical areas. It is native to the beaches and swamps of Southeast Asia.
Other names: English "Spider Lily," "Swamp Lily"; German "Weiße Hakenlilie."
Additional information: In the wild, the poison lily always grows close to bodies of water because its seeds are capable of floating and spreading on the stream or current. All parts of the plants are poisonous, but they are used for different medical purposes. The wild specimen has light green leaves and white flowers with a greenish tube. Cultivated specimens have lightly striped or violet leaves and flowers. The slightly smaller poison lily (*C. amabile*) always has green leaves, but its flowers are light red violet, and its buds are dark. *Hymenocallis littoralis*, the spider lily, looks very much like the poison lily, but a membranous limb connects the lower part of its stamens.

Heart of Jesus

Caladium bicolor
Family: *Araceae*

Most important features: The leaves are multicolored. Usually, they have white spots and red veins. At the bottom, the leaves are deeply incised, but not up to the stipula of the petiole.
Growth structure: The heart of Jesus is an herbaceous plant without an erect stem. All the leaves start from the ground and have a very long, firm petiole.
Leaves: The leaves are alternating and ovate to triangular in the outline. They are 4 to 15 inches (10 to 35 cm) long and 2.4 to 8 inches (6 to 20 cm) wide with three to seven firm veins running from the stipula. The lower pair runs into the basal end of the spherical lobes of the leaf and branches out again.
Flowers: The inflorescences have a greenish to cream white bract that creates an egg-shaped hull in the lower third. This encompasses a whitish to pale orange cob and stands off above it. The flowers themselves are tiny.
Fruits: The fruits are whitish berries with many seeds. They are densely packed on the cob.
Occurrence: Heart of Jesus grows as an ornamental plant in all tropical areas. The wild specimen originated in the northern part of South America.
Other names: English "Mother-in-Law Plant"; German "Buntblatt," "Buntwurz," "Kaladie"; French "Palette de Peintre"; Spanish "Capa de Rey," "Corazón de Jesús," "Corazón Sangrienta," "Hoja de Adorno," "Hocha de Leche."
Additional information: The heart of Jesus is a very decorative plant. One seldom sees its flowers or fruits. The wild specimen has red on the main veins and has white spots. However, numerous breedings have been created. These include the hybrid, *C. schomburgkii*. It has more distinctly triangular leaves with tapering lobes that are incised almost up to the stipula of the petiole.

Cayenne Pepper

Capsicum annuum
Family: *Solananceae*, Solanum

Most important features: Cayenne pepper is richly branched. Often it has two or three lateral branches next to each flower with small leaves in the axils of the larger ones. The flowers are greenish white to pale violet. They have five petals with violet stamens. The fruits are hollow with curved seeds.
Growth structure: The cayenne pepper is short and broad. It grows .2 to .6 inch (0.5 to 1.5 cm) high and is slightly woody at the bottom.
Leaves: The leaves are alternating, ovate to tapering, and thin. They are .6 to 4.8 inches (1.5 to 12 cm) long and .2 to 2.8 inches (0.5 to 7 cm) wide.
Flowers: The drooping flowers often grow individually. They are .6 to 1 inch (1.5 to 2.5 cm) long. The calyx is slightly dentate. At least half of the corolla is overgrown. The ovary has two to three chambers.
Fruits: The fruits come in a variety of sizes, shapes, and colors (see below). The septum is relatively thin. The seeds are .2 inch (5 mm) long and cream white.
Occurrence: This plant is cultivated worldwide. Originally, it was from the Western Hemisphere.
Other names: Numerous names, depending on the specimen (see below).
Additional information: The cultivated specimens of *Capsicum annuum* produce such a variety of fruits that one has a hard time recognizing them as the offspring of one single species. They range from small and very spicy, such as the chili specimens, to the oblong cayenne pepper, up to the large, spherical, mild bell pepper. These plants are cultivated almost all over the world as annuals, but they play a special role in the tropics because their spiciness stimulates the appetite, even in great heat. In addition, closely related species are called chili, especially *C. frutescens.* This is identified by its erect flowers and fruits, which grow in groups of two or three.

Boat Lily

Tradescantia spathacea
Family: *Commelinaceae*, Dayflower

Most important features: The leaves are oblong. On the top, they are dark green; on the bottom, they are red violet. The inflorescences are deep inside the axils of the leaves with a boat-shaped perianth in the bract.
Growth structure: The boat lily is a rosette plant that grows 10 to 24 inches (25 to 60 cm) high. The shoot is up to 8 inches (20 cm) long.
Leaves: The leaves are tapering, slightly thick, and often point upward. They are 6 to 18 inches (15 to 45 cm) long and 1 to 3.2 inches (2.5 to 8 cm) wide.
Flowers: The flowers are white with three small sepals. They are .4 to .6 inch (1 to 1.5 cm) long. The three petals and six stamens barely emerge from the 1 to 1.6 inch (2.5 to 4 cm) wide perianth of the bract. Only a few flowers are open at the same time.
Fruits: The fruits are capsules. They are elliptical and have two or three chambers. The fruits are about .2 inch (5 mm) long. They are enclosed in the perianth of the bract. Each chamber has one seed.
Occurrence: One finds the boat lily in gardens and on public grounds. It originated in Central America.
Other names: English "Moses-in-the-Cradle," "Oyster Plant"; German "Rhoeo," "Bootspflanze," "Wiegenlilie"; Portuguese "Roel."
Additional information: In earlier times, this species was called *Rhoeo discolor.* Most of the folk names refer to the boat or shell shape of the hull of the inflorescences. The plant is popular in the tropics as ground cover and as an indoor plant because it needs little care. Specimens with yellowish longitudinal stripes on the leaves are also cultivated. The closely related *T. pallida* gives an even more striking accent of color to many gardens. Its leaves are completely violet. Other *Tradescantia* species are also quite attractive, but they are often more sensitive.

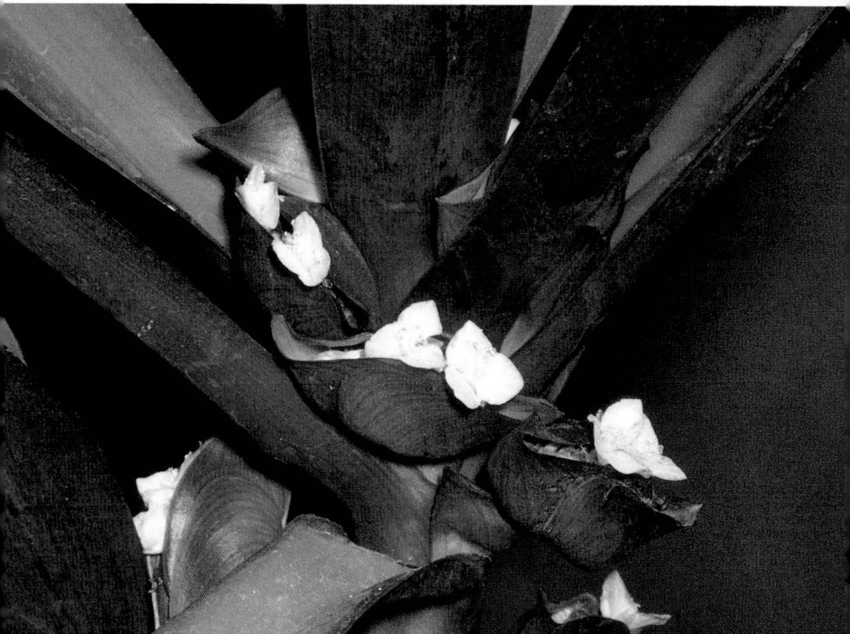

Blue Orchid

Vanda coerulea
Family: *Orchidaceae*, Orchids

Most important features: The flower has five petals with dark veins that form a pattern like a chessboard. The flower is often bluish. The three upper petals are slightly smaller than the two lower ones. A sixth petal is transformed into a short lip. This has a darker color and is almost straight after a bend at the base.
Growth structure: The plant grows erect. It is firm with thick aerial roots. The blue orchid grows up to 24 to 40 inches (60 to 100 cm) high.
Leaves: The leaves are arranged in two vertical rows. They are strap-shaped, thick, leathery, and blunt at the end. The leaves are 4.8 to 10 inches (12 to 25 cm) long and .6 to 1.2 inch (1.5 to 3 cm) wide.
Flowers: Five to twenty flowers grow in a simple raceme that is 12 to 20 inches (30 to 50 cm) long. The racemes usually emerge sideways from the axils. The flowers are 2.8 to 4.4 inches (7 to 11 cm) long. They are often light violet to azure blue. The petals have short petioles. They are tapering ovate to wide and spherical. The lip has two to three longitudinal folds and two small, white lateral lobes that are acute.
Fruits: The fruits are capsules with longitudinal slits and many tiny seeds.
Occurrence: Frequently cultivated, the blue orchid is native to the southern slope of the Himalayas from northeast India to the southern part of China and Southeast Asia. It is often found at altitudes of between 3,300 and 5,600 feet (1,000 and 1,700 m).
Other names: English "Blue Vanda"; German "Blaue Vanda."
Additional information: The blue orchid has become so rare in nature that it has the strictest injunction in *The Convention on International Trade in Endangered Species* (CITES). Without a proof of purchase, possession of a specimen is illegal. Although it is often cultivated because its blue flowers are very unusual for orchids, there are other color varieties, ranging from pink to slate gray. Between other *Vanda* species and the many breeding specimens, one can find almost all the colors of the rainbow (except green), often in combination with an interesting spotted pattern.

Christmas Orchid

Cattleya trianae
Family: *Orchidaceae*, Orchids

Most important features: The flowers are very large. The three outer petals are narrow; the two inner ones, wide and ovate, point toward the side. The third inner petal forms a tubular lip open at the top. The flower is crimson to purple, with a yellow to orange throat, and a wavy margin.
Growth structure: The plant grows erect with an oblong, thickened piece of stem that is often surrounded by a membranous hull under each leaf.
Leaves: The leaves are oblong, leathery, and slightly folded along the midrib. They are 6 to 10 inches (15 to 25 cm) long and 1.6 to 3.2 inches (4 to 8 cm) wide.
Flowers: Each stem has one to four flowers. The wild specimen is entirely pale pink. It is 6 to 7.2 inches (15 to 18 cm) long with a midcolumn hidden in the lip.
Fruits: The fruits are ovate to tapering with strongly emerging longitudinal ribs.
Occurrence: The Christmas orchid is frequently cultivated. It also often grows on trees. It was originally from Colombia.
Other name: English "Winter Cattleya"; German "Weihnachtsorchidee."
Additional information: The Christmas orchid's name is based on the fact that it flowers in greenhouses from December to February. It is the national flower of Colombia, yet it seldom grows wild there. Numerous color varieties are cultivated. These range from plain white to yellowish and red shades up to lavender blue. They are usually the starting specimens when breeding *Cattleya* hybrids. The hybrids are often unnaturally large and garishly colored. Another important parent is *C. bowringiana* (see photo), whose flowers are only half the size of the Christmas orchid, but whose petioles have forty-five leaves that are brighter in color. The closely related *Epidendrum* (see page 248), *Laelia*, and *Sophronotis* are frequently used to breed new specimens.

Fiery Reed Orchid

Epidendrum ibaguense
Family: *Orchidaceae*, Orchids

Most important features: The stem is long and thin. The leaves are arranged in two vertical rows. The flowers are dark red to almost yellow with five petals and a lip that is attached to the midcolumn. The flower ends in three lobes that are fringed.
Growth structure: The stem is erect and can climb. It is often about 3 feet (1 m) long and only .2 inch (5 mm) thick. Much of the stem is covered by brownish leaf sheathes.
Leaves: The leaves are sessile, ovate to tapering, and fleshy. They are 1.4 to 6.4 inches (3.5 to 16 cm) long and .4 to 2 inches (1 to 5 cm) wide.
Flowers: The flowers grow in spherical or dense simple racemes that point upward. The racemes grow up to 6 inches (15 cm) long. The stem has only scale leaves on the last 4 to 20 inches (10 to 50 cm). The petals are quite similar to one another. They are .6 to .8 inch (15 to 20 mm) long and .2 to .3 inch (5 to 8 mm) wide. The lip is .3 to .6 (8 to 15 mm) long.
Fruits: The fruits are crooked to elliptical. They are 1 to 1.6 inch (2.5 to 4 cm) long with many tiny seeds.
Occurrence: The plant is widely spread in Central and South America. One often sees it growing on rocks. It also grows on trees, but less often. In other locations, one only sees it in cultivation.
Other names: English "Reed Stem Epidendrum"; German "Feuerorchidee"; Spanish "Boca de Fuego," "Rancho Viejo."
Additional information: The fiery reed orchid can grow almost anywhere, on plain rocks as well as on sandy areas, on trees, or even on floating islands created by flotsam and water plants. Often, one finds them on wasteland together with silkweed and cherry pies (see page 142), whose flowers have similar colors. If their stems touch the ground, they often develop new roots at the knots and continue to grow. In this way, their stem can grow up to 33 feet (10 m) long. Many of the approximately eight hundred *Epidendrum* species are also kept as ornamental plants, for example *E. magnificum* (see photo at right).

Dancing Doll Orchid

Oncidium flexuosum
Family: *Orchidaceae*, Orchids

Most important features: The flowers are .6 to .8 inch (1.5 to 2 cm) long with five small, tapering petals. The petals are yellow and brown, spotted or banded, and often curled with a bright yellow lip and two small lateral lobes. They have a large, wide, and spherical terminal lobe that is deeply scalloped at the end.
Growth structure: This orchid grows with thickened, flattened shoots that are .8 to 3.2 inches (2 to 8 cm) long and appear at intervals of .8 to 2 inches (2 to 5 cm) on a climbing rootstock.
Leaves: The leaves grow individually or in pairs on each segment of the shoot. The leaves are 4 to 15 inches (10 to 35 cm) long, 1 to 2 inches (2.5 to 5 cm) wide, and tapering.
Flowers: Numerous flowers grow on panicles that are 24 to 48 inches (60 to 120 cm) long. The petals are frizzy on the margin. The lip is smooth. At the bottom of the lip is a fleshy formation with three parts. It is partly light or reddish and ribbed.
Fruits: The fruits are capsules that open along longitudinal slits, exposing many tiny seeds.
Occurrence: The dancing doll orchid is widely cultivated. Originally, it was from Brazil.
Other names: German "Oncidium"; Portuguese "Chuva de Oro."
Additional information: The seven hundred species of the *Oncidium* genus are represented almost everywhere in the Western Hemisphere. The flowers of most species are bright yellow and shiny brown. Actually, these colors are rather unusual for orchids. In some cases, however, reddish or violet shades dominate. In the wind, male bees sometimes push the dancing panicles of *Oncidium*. To the bees, the flowers may represent a rival. Female bees of the same species behave completely differently. They sit peacefully on the lip and collect oil to feed their larva.

Dendrobium

Dendrobium nobile
Family: *Orchidaceae*, Orchids

Most important features: The petals are almost entirely white at the bottom. At the tip, they are violet pink. The three outer ones are narrow; the two inner ones are a little wider. They are also elliptical and point to the side or slightly upward. The third inner petal is transformed into a cone-shaped lip that is open at the top. At the tip, it is violet pink. In the middle, it is white to yellowish; in the throat, it is velvet wine red to purple brown.
Growth structure: Dendrobium grows erect with several fleshy, grooved stems that usually grow in segments surrounded by a membranous hull. The stems are 12 to 28 inches (30 to 75 cm) long.
Leaves: The leaves grow in two vertical rows. They are oblong to elliptical with a crooked end. The leaves are 2.8 to 4.4 inches (7 to 11 cm) long and .8 to 1.2 inch (2 to 3 cm) wide.
Flowers: The flowers are aromatic and usually grow in numerous inflorescences on long, leafless stems that each have two to four flowers. The flowers are 2 to 3.2 inches (5 to 8 cm) long with a midcolumn hidden in the lip.
Fruits: The fruits are short, squared capsules with many tiny seeds.
Occurrence: This orchid is frequently cultivated. Originally, it spread from the northern part of India to southern China and Thailand.
Other names: This orchid is known throughout the world by its scientific name.
Additional information: *Dendrobium nobile* is the national flower of Sikkim. It thrives in altitudes up to 5,900 feet (1,800 m) and needs cool nights in order to blossom. Large specimens can have two to three hundred flowers at the same time. Many of the more than nine hundred *Dendrobium* species are kept as ornamental plants. The usually red flowers of *D. phalaenopsis* are similar in shape to the *Phalaenopsis* genus (see below), but they have a cone-shaped, acute lip. The flowers of *D. thrysiflorum* hang in dense simple racemes and are white with a yellow lip.

Butterfly Orchid

Phalaenopsis amabilis
Family: *Orchidaceae*, Orchids

Most important features: The flowers are white with three narrow petals and two wide petals. The sixth is transformed into a lip. The flowers grow on petioles. At the bottom, are two large, lateral lobes that are bent upward. The lobes are yellow with red spots and a firm double lump in the middle. The petals have two filamentlike appendages.
Growth structure: The butterfly orchid often has only two to four leaves on a short shoot with thick, greenish white aerial roots.
Leaves: The leaves are tapering or slightly wider above the middle. They are 8 to 12 inches (20 to 30 cm) long and 1.6 to 4.8 inches (4 to 12 cm) wide. The leaves are fleshy and often bent toward the ground. Usually, they are rounded at the end and slightly reddish at the bottom.
Flowers: The flowers grow in groups of six to fifteen in drooping or hanging panicles that are 12 to 40 inches (30 to 100 cm) long. The flowers are 3.2 to 4.8 inches (8 to 12 cm) long. The outer petals are narrow and elliptical; the two lower ones are often slightly crooked. The inner ones face sideways. The flowers grow on short petioles.
Fruits: The fruits are shaped like bars. They grow up to 2.4 inches (6 cm) long and have many tiny seeds.
Occurrence: This is one of the most frequently cultivated orchids. Originally, it was from the Indo-Malaysian islands.
Other names: English "Moth Orchid"; German "Mondorchidee," "Falterorchidee," "Malaienblume."
Additional information: The butterfly orchid is one of the three national flowers of Indonesia. Its name refers to the shape and color of the flowers and to its durability; it lasts through an entire phase of the moon. Many of the approximately forty-five *Phalaenopsis* species are cultivated. They are also used to breed new specimens, particularly the pink *Ph. schilleriana* (see photo at right).

Honey Plant

Hoya carnosa
Family: *Axlepiadaceae*, Silkweed

Most important features: The leaves are thick and fleshy. They are set opposite on thin, climbing shoots. The flowers grow in dense simple umbels. They are whitish and reddish with five petals arranged in the shape of a star.
Growth structure: The honey plant grows as a thin shoot up to 20 feet (6 m) long. It has short roots that adhere to trees and rocks.
Leaves: The leaves are ovate to elliptical, waxy, shiny, spherical to slightly heart-shaped on the bottom, and sometimes pointed. They are 2 to 4 inches (5 to 10 cm) long and 1 to 2.2 inches (2.5 to 5.5 cm) wide.
Flowers: The flowers grow on petioles that are .4 to 1.2 inch (1 to 3 cm) long. The simple umbels have up to thirty flowers. These are .8 to 1.6 inch (2 to 4 cm) long and .6 to .8 inch (1.5 to 2 cm) wide. The sepals are up to .2 inch (5 mm) long and hidden under the corolla, which is velvety and white to delicate pink. Above it is a smaller corona, which is star-shaped, shiny, cream white, and red in the middle.
Fruits: The fruits are tapering. They are 2.4 to 4 inches (6 to 10 cm) long and .2 to .6 inch (0.5 to 1.5 cm) thick.
Occurrence: One finds the honey plant on trees, rocks, and occasionally on white sand. Originally, it spread from India to China and Australia. It is used as an ornamental plant in other areas as well.
Other names: English "Porcelain Flower," "Wax Plant"; German "Wachsblume," "Porzellanblume"; Portuguese and Spanish "Flor de Cera."
Additional information: The honey plant is the most popular of the approximately seventy *Hoya* species. Many of these grow on trees in Southeast Asia. Most of them have a somewhat artificial appearance, as if they were made of wax or porcelain. The name refers to the fact that large drops of nectar often collect on the corona. There, they attract nocturnal insects. The nectar gives the flowers a pleasant aroma in the evening. In the Western Hemisphere, hummingbirds also drink the nectar.

Lipstick Plant

Aeschynanthus radicans
Family: *Gesnericaceae*, Gesneria

Most important features: The leaves are set opposite. The flowers are bright red with a long tube, yellow spots in the throat, and five lobes at the end. The two upper ones are stunted; they are less than half of the length of the others.
Growth structure: This is a crawling plant that grows up to 5 feet (1.5 m) long. All parts of the plant are hairy.
Leaves: The leaves are ovate to tapering and elliptical. They are .8 to 1.8 inch (2 to 4.5 cm) long and .4 to 1 inch (1 to 2.5 cm) wide. The leaves are slightly fleshy.
Flowers: The flowers grow individually in the axils or accumulate at the ends of the stems. The calyx is green to dark purple brown and tubular. It is .8 to 1 inch (2 to 2.5 cm) long and has five short, relatively rounded cusps at the end. The corolla tube is 1.8 to 2.4 inches (4.5 to 6 cm) long and slightly crooked. The pistil and four stamens (two longer and two shorter ones) are situated below the labrum.
Fruits: The fruits are tapering capsules. They are 1 to 1.6 inch (2.5 to 4 cm) long. When they are ripe, the two flaps open.
Occurrence: The lipstick plant grows wild on trees and rocks in Southeast Asia. In other areas, it is only grown as an ornamental plant in a pot.
Other names: German "Lippenstiftblume," "Schamblume."
Additional information: One finds representatives of the 140 species of the *Aeschynanthus* genus everywhere in Southeast Asia, especially in the mountain forests. Although some thrive on the floor, the majority grow on trees. Its bright red flowers indicate that it is pollinated by birds. In the tropical areas of the Western Hemisphere, the *Columnea* genus, with its seventy-five species, is similar, although it grows berries. Many other of these family members are popular as ornamental plants, such as the African violet (*Saintpaulia*), the slipper gloxinia (*Sinningia*), and the cape primrose (*Streptocarpus*).

Silver Vase

Aechmea fasciata
Family: *Bromeliaceae*

Most important features: The leaves are dark green. They may have light gray spots on both sides or only on the outside. They also have cross stripes. The leaves create a funnel-shaped cistern. The inflorescence is pink with many slender, tapering, dentate bracts. At first, the flowers are pink; later, they are blue.
Growth structure: The plant grows as a funnel-shaped rosette with ten to twenty leaves.
Leaves: The leaves are wide bands that rise steeply and then often droop at the tip. They are 12 to 40 inches (30 to 100 cm) long and 1.2 to 3.2 inches (3 to 8 cm) wide. The leaves are very stiff, and the margins have very sharp thorns. **Flowers:** The flowers have three petals. They are up to 1.4 inch (3.5 cm) long with an inferior ovary cramped between the much longer and densely arranged bracts of the inflorescence, which is 2.4 to 3.2 inches (6 to 8 cm) long. The petiole is up to 12 inches (30 cm) long.
Fruits: The fruits are berries, crowned by the remainders of the perianth and hidden between the bracts of the inflorescence.
Occurrence: The silver vase is popular as an ornamental plant. It is also grown in a pot. Originally, it was from Brazil.
Other names: English "Urn Plant"; German "Aechmea," "Lanzenrosette"; Portuguese "Aequiméia."
Additional information: The silver vase is one of more than 2,400 species of the Bromeliaceae family. Most of these grow on trees. The family shows an impressive variety of flowers (see photo at right, *Guzmania sanguinea*); however, the growth structures are often very similar. The cistern created by the base of the leaf collects leaf litter and rainwater; thus the plant doesn't need the ground.

Bird's Nest Fern

Asplenium nidus
Family: *Aspleniaceae*, Spleenwort

Most important features: The leaves often have diagonal brown stripes on the bottom. This creates a funnel-shaped rosette, the wide center of which is flatly arch-shaped and covered densely with almost black scales.
Growth structure: The plant grows as a rosette with diagonal, erect, fresh leaves. Often it also has hanging old leaves.
Leaves: The leaves are tapering with indistinct petioles, a wavy margin, an almost black midrib, and delicate lateral veins that are only connected to each other on the margin. The leaves are 12 to 60 inches (30 to 150 cm) long and 1.6 to 8 inches (4 to 20 cm) wide. They are widest above the middle. The youngest leaves are curled.
Spore reservoirs: These are recognizable as diagonal strokes on the bottom of some leaves. The reservoirs are .6 to 1.6 inch (1.5 to 4 cm) long. They are closer to the midrib than to the margin.
Occurrence: One finds this plant growing in the shade at the top of trees. It is native to the tropics of the Eastern Hemisphere.
Other names: German "Nestfarn"; French "Fougère Nid-d'oiseau"; Spanish "Helecho Nido."
Additional information: The bird's nest fern only needs a little light in order to grow. As a result, it is very popular as an indoor plant. However, low humidity will stunt the young leaves. The bird's nest fern is different from many other Bromelias (see above) because its leaf funnel is not dense enough to retain water. However, over time, it collects a large amount of leaf litter. The plant draws nutrients and moisture from this. Many small roots grow from the shoot into the funnel. Some ferns also have suction scales on its crawling shoot axes, yet not on its leaves (compare with page 258).

Antler Fern

Platycerium coronarium
Family: *Polypodiaceae*, Polypody

Most important features: This plant, which can be quite large, grows on trees. It has wide nest leaves that are sessile, diagonal, erect, and lobed. The leaves may also be hanging, multiply branched, deciduous leaves.
Growth structure: Often the antler fern grows as a semirosette in the forks of branches or on the side of the trunk.
Leaves: The nest leaves are 20 to 40 inches (50 to 100 cm) long and up to 10 inches (25 cm) wide. They are deeply lobed with strongly emerging, branched veins. The bottom is wide, heart-shaped, and fleshy. The deciduous leaves have one very long lobe at the bottom and several shorter lobes. The long lobe is 3 to 13 feet (1 to 4 m) long with bandlike segments and .8 to 1.6 inch (2 to 4 cm) wide. Often, it is twisted. At the bottom, hidden under the short lobes, is a semispherical to kidney-shaped formation that is up to 10 inches (25 cm) wide. The concave bottom of this formation is entirely covered by spore reservoirs.
Spore reservoirs: These are brown, feltlike, and located on the spherical lobe of the deciduous leaves (see above).
Occurrence: Originally spread from Myanmar to the Philippines, the antler fern is now seen as an ornamental plant in parks.
Other names: English "Crown Staghorn Fern"; German "Kronen-Geweihfarn."
Additional information: Young plants only have nest leaves in the beginning. They remain green for a long time and develop a large basket in which leaf litter accumulates. Later on, the nest leaves turn brown, bend inward, and thus secure the compost hidden beneath them. The deciduous leaves that carry spores grow later on. The deciduous leaves of *P. superbum* have a large spore reservoir at its first branching. With the smaller common staghorn fern (*P. bifurcatum*), which is also kept as an indoor plant, the spore reservoir covers the bottom side at the tips of the lobes.

Oak Leaf Fern

Drynaria quercifolia
Family: *Polypodiaceae*, Polypody

Most important features: This is a fern with long, deeply pinnately parted fronds and many smaller, sessile nest leaves. These are usually brown and similar to oak leaves.
Growth structure: The oak leaf fern often grows as a semirosette in the fork of branches or an axis of the shoot. It rises on the tree. It is about .8 inch (2 cm) thick. The axis of the shoot is densely covered by long, brown scales.
Leaves: The leaves are erect, slightly drooping, and deciduous. They are 12 to 52 inches (30 to 130 cm) long. The pinnas are .8 to 12 inches (2 to 30 cm) long and .4 to 2 inches (1 to 5 cm) wide. They are tapering, widely sessile, and predominantly connected along the midrib. The nest leaves are 2.4 to 16 inches (6 to 40 cm) long and 3.6 to 12 inches (9 to 30 cm) wide. The indentations may be .8 to 2 inches (2 to 5 cm) deep.
Spore reservoirs: These are tiny and round. They grow on the bottom of some deciduous leaves in two regular rows between the larger lateral veins of the pinnas.
Occurrence: Originally, the oak leaf fern spread from India to Fiji and across all of tropical Asia. Today, one can find it in many other areas, too.
Other name: German "Eichenblattfern."
Additional information: The nest leaves of the oak leaf fern are green for only a short time; however, they remain on the plant for a very long period. Leaf litter accumulates between the leaves. This rapidly decomposes in the humid and warm climate, providing a supply of nutrients for the plant. In addition, ants may populate the niches and provide additional nutrients. The closely related species, *D. sparsisora*, is also known as an oak leaf fern. It differs from *D. quercifolia* because its spore reservoirs are irregularly spaced. *D. rigidula* has pinnas with petioles and hanging fronds.

Louisiana Moss

Tillandsia usneoides
Family: *Bromeliaceae*

Most important features: Louisiana moss is a gray, hanging lichen. When closely examined, it is greenish and covered with tiny, silvery, little scales. The green is more noticeable when the plant is moistened.
Growth structure: This hanging plant grows with tiny filaments. It is up to 26 feet (8 m) long and richly branched.
Leaves: The leaves are filaments that are .6 to 2.4 inches (1.5 to 6 cm) long. They are slightly flattened and gradually tapering. They are difficult to distinguish from the shoots.
Flowers: The flowers grow individually. The sepals are .3 inch (8 mm) long. They are membraneous and hardly recognizable as sepals. The petals are greenish yellow to pale bluish and .3 inch (8 mm) long. The last ones are .2 (5 mm) long and spread out or curled.
Fruits: The fruits are cylindrical capsules with a small pointed end. They are up to 1 inch (2.5 cm) long.
Occurrence: Louisiana moss grows on trees, rocks, houses, and even overhead power lines in all warm areas of the Western Hemisphere from the southeastern part of the United States to Argentina.
Other names: English "Old Man's Beard," "Spanish Moss"; German "Louisianamoos"; Portuguese "Barba de Pau, "Barba de Velho"; Spanish "Barba de Viejo," "Musgo Blanco."
Additional information: Louisiana moss is not a moss. It is recognizable as Bromelia (compare with page 254) because of its flowers. Only seedlings have roots; later on, the plant provides itself with water and nutrients from the silvery, water-absorbing peltate scales on its leaves and shoots. As far back as the Aztecs, it was used to decorate temples. In Central America, this custom was adapted to Christian altars. Today, the plant is used in flower arrangements and as packing material.

Mistletoe Cactus

Rhipsalis baccifera
Family: *Cactaceae*, Cacti

Most important features: The mistletoe cactus is shaped like a bar. It has hanging shoots and regular spots, but no leaves. It often has small yellowish or greenish white flowers or white to pink berries.
Growth structure: The shoot hangs from the tree. It is 1.5 to 10 feet (0.5 to 3 m) long and .2 to .3 inch (5 to 8 mm) thick. At intervals of 2 to 12 inches (5 to 30 cm), it is whorled and branched, sometimes with short aerial roots.
Leaves: At most, these are recognizable as tiny scales.
Flowers: The flower is sessile and up to .2 inch (5 mm) long. Half of it consists of the inferior ovary. It has five tiny sepals. The flower also has five oval petals that are .2 (5 mm) long. It has many yellow stamens and a pistil with three to five stigmas.
Fruits: The fruits are spherical or slightly tapering. They are .2 to .3 inch (5 to 8 mm) long and crowned by the remainders of the perianth. They are transparent. Inside are many black seeds in a very sticky juice.
Occurrence: The mistletoe cactus is widely spread from the Western Hemisphere to Africa and Madagascar as well as Sri Lanka. It is seldom cultivated.
Other names: German "Mistelkaktus."
Additional information: At first sight, it's difficult to identify the mistletoe cactus as a cactus. One needs to look at the small points (areolas) on the stem and at the flowers to be sure. Young plants are easier to identify because they still have squared stems with .2 to .3 inch (5 to 8 mm) long thorns. The mistletoe part of the name is based on the fact that the plant lives on trees. In addition, the whitish, sticky fruits are similar to the fruits of the mistletoe. This is the only cactus that is native to the Western Hemisphere. It differs from the mistletoe in that it is not a parasite; it simply uses its host as an understock.

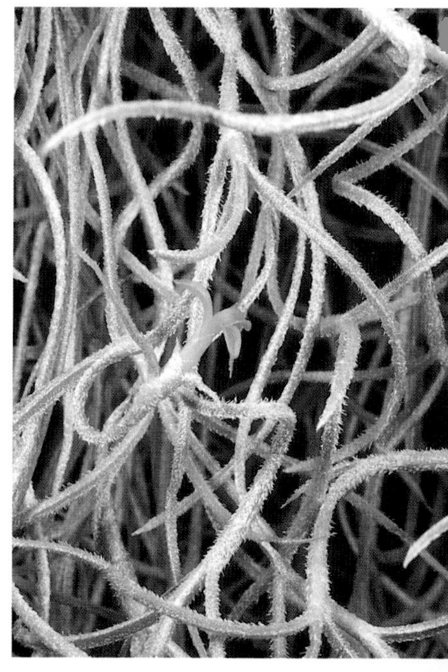

Water Hyacinth

Eichhornia crassipes
Family: *Pontederiaceae*, Pontederia

Most important features: The base of the petioles is swollen and keeps much of the plant floating on top of the water. The flowers are light blue to violet. The highest of the six petals has a yellow spot enclosed in a darker area.
Growth structure: These are freely swimming rosettes that are seldom anchored in the mud. They are 8 to 20 inches (20 to 50 cm) high with long, hairy roots.
Leaves: The leaves are green, shiny, and spherical to kidney-shaped. They are 2 to 6 inches (5 to 15 cm) long with very delicate, basal veins.
Flowers: The flowers grow in inflorescences that are similar to panicles. They are 1.6 to 15 inches (4 to 35 cm) high. The flowers are 1.6 to 2.4 inches (4 to 6 cm) long with three short and three long stamens that are bent upward.
Fruits: The fruits are green, wrinkly capsules hidden between the remainders of the perianth. They are about .6 inch (1.5 cm) long with many slender winged seeds.
Occurrence: One finds these plants in standing and slowly running water everywhere in the tropics. Originally, they were from South America.
Other names: German "Wasserhyazinthe"; French "Jacinthe d'Eau"; Portuguese "Aguapé," "Jacinto Aquático"; Spanish "Buchón de Agua," "Camalote," "Lírio Aquático."
Additional information: The water hyacinth was introduced into all tropical areas because of its beautiful flowers; however, in many areas, it grew out of control. In addition to reproducing with its seeds, it reproduces with runners, and almost every segment can grow into a plant. In this way, a single rosette is able to produce more than three thousand descendents within a year. In areas where it lacks natural enemies, the water hyacinth often develops such dense mats that it clogs waterways, even hindering the passage of ships. On the other hand, in Southeast Asia, it serves as an almost inexhaustible source of food for pigs.

Santa Cruz Water Lily

Victoria cruziana
Family: *Nymphaeaceae*, Water lilies

Most important features: The floating leaves are very large. The margins are bent upward. The flowers, too, are very large. At first, they are white; later on, they are dark pink.
Growth structure: This is a floating leaf plant with a thick rootstock anchored in the mud.
Leaves: The leaves are circles. The bottom has firm ribs and thorns. The leaves are 20 to 60 inches (50 to 150 cm) wide, and the margin is 3.2 to 8 inches (8 to 20 cm) high.
Flowers: The flowers grow individually on the surface of the water. They are 10 to 15 inches (25 to 35 cm) wide with four thorny sepals at the bottom. The many petals and stamens blend into one another and into the bowl-shaped stigma, which has bent appendages on the margin.
Fruits: The fruits ripen under water.
Occurrence: The Santa Cruz water lily is widely cultivated in standing bodies of water. It is native to Paraguay, the southern part of Brazil, and the northern part of Argentina.
Other names: German "Paraná-Riesenseerose," "Irupé."
Additional information: The Santa Cruz water lily is not quite as famous as the Amazon water lily (*V. amazonica*), but it is cultivated more frequently. It thrives in water temperatures that range from 68° to 77° F (20° to 25° C). *V. amazonica*, on the other hand, needs temperatures of 80° to 86° F (27° to 30° C). Even in the tropics, the water may not always be that warm. The leaves of *V. amazonica* grow up to 7 feet (2 m) long. However, the margin is only 1.6 to 4 inches (4 to 10 cm) high. The leaves of both species are strong enough to support a small child. Photos showing adults on the leaves are always faked in some way. A crossbreed of the two species is cultivated. In addition, one can find numerous species of the real water lilies (*Nymphaea*) in the tropics. The blue flowering *N. caerulea* from Africa and the crimson *N. rubra* from India with their breeding specimens are particularly popular.

Sacred Lotus

Nelumbo nucifera
Family: *Nelumbonaeceae*, Butterjags

Most important features: The leaves are almost circular. The stipula of the petiole is close to the middle. The flowers are large with many pink (occasionally white) petals and yellow stamens. In the middle is a yellow cone that stands upside down. It has little holes on the upper side.
Growth structure: The sacred lotus grows with a firm crawling rootstock. The plant grows in swamps with the leaves and flowers elevated so that they are out of the water.
Leaves: The older leaves are waxy and grow on long petioles that are up to 5 feet (1.5 m) above the water. The leaves are flat to cone-shaped and 12 to 35 inches (30 to 90 cm) wide.
Flowers: The flowers grow individually. Often they stand slightly farther above the water than the leaves. The flowers are 4.8 to 12 inches (12 to 30 cm) long and have a pleasant aroma.
Fruits: The fruits are elliptical and often bluish. They are .6 to 1 inch (1.5 to 2.5 cm) long and grow individually in the hollow spaces of the flower cones that are 2 to 4.8 inches (5 to 12 cm) wide. The flower cones are green to brown when the fruits are ripe.
Occurrence: One finds the sacred lotus in standing water that does not freeze all the way to the bottom. The plant is native to Asia, Japan, and even to Australia.
Other names: German "Lotusblume."
Additional information: This plant is sacred to Hindus and Buddhists as a symbol of rebirth and purity. However, it also has secular uses. For example, people eat the rootstock, young leaves, and the fruits. Although the sacred lotus grows in muddy ponds, it is always very clean. Small wax structures on its surface, which can only be seen under a microscope, prevent dirt from adhering. The dirt is always washed off with the next rainfall.

Sawah Flowering Rush

Limnocharis flava
Family: *Alismataceae*, Water plantain

Most important features: The stem and petioles have three sides. The leaves have arch-shaped running veins. The petioles are thicker than the buds. The flowers have three green sepals and three yellow petals with many stamens and carpels.
Growth structure: The plant grows as erect rosettes from a crawling rootstock. It grows 12 to 44 inches (30 to 110 cm) high.
Leaves: The leaves are dull and relatively pale green. They grow on petioles that are 4 to 32 inches (10 to 85 cm) long. The elliptical leaf blade is 2.8 to 11 inches (7 to 28 cm) long and 1.2 to 8.4 inches (3 to 21 cm) wide. The leaf is spherical to heart-shaped on the bottom.
Flowers: The flowers grow in groups of three to twelve in one simple umbel at the end of a stem that is 10 to 28 inches (25 to 75 cm) long. The petiole is 1.2 to 3.2 inches (3 to 8 cm) long. The sepals are about .6 inch (1.5 cm) long, and the petals are .8 to 1 inch (2 to 2.5 cm) long.
Fruits: The fruits are .4 to .6 inch (1 to 1.5 cm) long. Each petiole has fifteen to twenty semicircular fruits.
Occurrence: The plant grows in muddy meadows, in ditches, and in rice fields. Originally, it was from South America.
Other names: English "Sawah Lettuce," "Velvetleaf," Yellow Burhead"; German "Gelber Froschlöffel"; Spanish "Cebolla de Chucho," "Hoja de Buitre."
Additional information: Like most water plants, the sawah flowering rush grows very rapidly, and it becomes a problem. In Southeast Asia, people eat the young leaves and inflorescences as vegetables. The entire plant is also used to feed pigs. The sawah flowering rush is considered dangerous because at least one of the most important herbicides has no effect on it at all. Thus, one cannot legally import it into the United States.

Water Lettuce

Pistia stratiotes
Family: *Araceae*

Most important features: The water lettuce consists of freely floating rosettes growing from sessile leaves that are soft and hairy. The leaves are dull and relatively pale green. They have three to twelve parallel running veins that are immersed at the top and strongly emerging at the bottom.
Growth structure: Stretched out or erect, the plant is up to 10 inches (25 cm) wide. It grows with runners and many delicately pinnate roots.
Leaves: The leaves are wide and shaped like wedges. They are spherical at the end and may be slightly scalloped. The leaves are .8 to 6 inches (2 to 15 cm) long.
Flowers: The flowers are tiny. Each has one female and a few male flowers in a .4 to 1.2 inch (1 to 3 cm) long bract that is greenish white and cone-shaped. The bract is pulled together in the middle. The flowers are hidden between the leaves and rarely seen.
Fruits: The berries are egg-shaped and about .2 inch (5 mm) thick. They are seldom seen.
Occurrence: Water lettuce grows in standing and slowly flowing bodies of water. The place of origin is uncertain.
Other names: German "Wassersalat," "Muschelblume"; French "Laitue d'Eau"; Portuguese "Alface d'Agua," "Mururé Pajé"; Spanish "Lechuga de Agua," "Repollito de Agua."
Additional information: The water lettuce reproduces primarily with runners. It reproduces so rapidly that descendents of one plant can cover large areas of water. The hair on the leaves does not become wet. A great deal of air is trapped between the hairs, giving the plant buoyancy. For this reason, even large water birds can walk on an area covered by water lettuce. Traditionally, it is used as food for ducks and pigs. The water lettuce is cultivated in aquariums and in ornamental ponds in summer, but it can quickly grow out of control.

Bulrush

Cyperus papyrus
Family: *Cyperaceae*

Most important features: The stem is dull and three sided. It grows 8.5 feet (2.5 m) high. At the top, the stem is covered with an umbrella-shaped bunch of thin, green branches that carry spikes.
Growth structure: The plant grows erect on a firm rootstock that crawls in the bottom of the swamp. The leaf sheathes are up to 20 inches (50 cm) long.
Leaves: The green leaves only grow on the stems that don't flower. They are grasslike and .3 inch (8 mm) wide. Otherwise, the leaves are only delicate bracts in the inflorescences.
Flowers: The flowers are tiny and grow in small spikes consisting of twelve to forty flowers. The spikes are .6 to 1.2 inch (1.5 to 3 cm) long and grow in groups of two to five on the numerous (200 to 360) rays of the inflorescences. The inflorescences are 2 to 16 inches (5 to 40 cm) long. They grow at the top of the stem. The bottom of the inflorescence has several scale leaves that are light brown. The scale leaves are 2 to 4 inches (5 to 10 cm) long and .4 to 1.2 inch (1 to 3 cm) wide.
Fruits: The fruits are tapering, three-sided, and tiny with dry membranes.
Occurrence: One finds this plant in calm bodies of water and near embankments. Originally, it was from Africa.
Other names: English "Papyrus," "Egyptian Paper Plant"; German "Papyrus"; French "Papier du Nil"; Portuguese and Spanish also "Papiro."
Additional information: This is the plant from which Egyptians produced a sort of paper over five thousand years ago. They cut the core of the stem into strips and pressed many strips firmly together until they dried. Bulrush was also used to manufacture baskets, shoes, and boats. Several other *Cyperus* species are also cultivated, but these are smaller plants. For example, the umbrella plant (*C. involucratus*) is cultivated as an indoor plant.

Gumbo

Abelmoschus esculentus
Family: *Malvaceae*, Mallow plants

Most important features: The leaves are digitately arranged or lobed. The flowers are large and yellow. In the center, they are dark red with a midcolumn formed from many yellow stamens. At the end are five dark red stigmas. The fruits are oblong and erect. They have five to nine ribs that run lengthwise.
Growth structure: Gumbo grows erect as a firm herb. It is often slightly branched and grows 3 to 13 feet (1 to 4 m) high.
Leaves: The leaves are alternating and grow on a petiole that is 2 to 20 inches (5 to 50 cm) long. The leaf blade is 4 to 15 inches (10 to 35 cm) long and just as wide. It is slightly to deeply lobed and serrated on the margin.
Flowers: The flowers grow individually in the axils. They are shortly sessile with an epicalyx that consists of eight to fifteen slender, small leaves that are .4 to 1 inch (1 to 2.5 cm) long. The calyx is .6 to 1.8 inch (1.5 to 4.5 cm) long, dentate, and split on one side. The petals are wide and asymmetrical. They are 1.2 to 2.4 inches (3 to 6 cm) long and spherical at the end.
Fruits: The fruits are green for a long time. When they ripen, they turn brown. The fruits are 3.2 to 12 inches (8 to 30 cm) long and .6 to 1.2 inch (1.5 to 3 cm) thick. They are tapered and slightly thickened at the bottom.
Occurrence: Gumbo is cultivated everywhere in the tropics and subtropics. It originated in the Eastern Hemisphere, but the exact location is unknown.
Other names: English "Lady's Fingers"; German "Okra"; French "Gombo"; Portuguese "Quiabeiro"; Spanish also "Quimbombo."
Additional information: Because gumbo is an annual plant, it usually dies after its fruits ripen. People often cook the unripe fruits and eat them as vegetables. However, they can also be eaten uncooked. They have a high grease content. *A. moschatus* is a similar species with stronger hair growth, larger flowers, and ovate fruits. It is cultivated by the perfume industry for its fragrant seeds.

Egyptian Cotton

Gossypium barbadense
Family: *Malvaceae*, Mallow plants

Most important features: All parts of the plant are green with small, dark points. Most of the leaves have three lobes, but a few have no lobes, and some have five lobes. The flowers have three large epicalyx bracts around the cup-shaped calyx. These have an almost continuous margin. The five petals are yellow at the bottom. Each petal has a purple spot. As the flower wilts, this spot becomes pink or red.
Growth structure: Egyptian cotton grows as a bush with a few woody branches. It grows 3 to 13 feet (1 to 4 m) high.
Leaves: The stipules and petioles are shed early. The leaf blade is 2 to 6 inches (5 to 15 cm) long and just as wide with acute lobes. The lobes on the side are usually smaller than the one in the middle. The lobes have five main veins. They start from the heart-shaped bottom.
Flowers: The flowers grow individually in the axils. The corolla is funnel-shaped. The petals are twisted like the blades of a propeller. They are 1.4 to 2.2 inches (3.5 to 5.5 cm) long. The midcolumn has many stamens and three stigmas.
Fruits: The fruits are ovate and tapering with three chambers. They are 1.2 to 2.4 inches (3 to 6 cm) long. The cotton is easy to detach from the seeds.
Occurrence: Cultivated everywhere in warm areas, Egyptian cotton is often grown on large fields. It originated in the Western Hemisphere.
Other names: English "Pima Cotton," "Sea Island Cotton"; German "Westindische Baumwolle"; Spanish "Algodon del País."
Additional information: Historians believe that Egyptian cotton was used in Peru ten thousand years ago. This would make it one of the oldest known cultural plants. Four of the approximately forty *Gossypium* species are used for cotton. *G. Hisutum* comes from the Western Hemisphere. It has slightly smaller, paler flowers and covers the largest fields of cultivation in the subtropics. *G. arboreum* is very similar, but it comes from the Eastern Hemisphere. The smaller, annual, *G. Herbaceum*, is cultivated less often.

Vanilla

Vanilla planifolia
Family: *Orchidaceae*, Orchids

Most important features: Vanilla is a climbing plant with fleshy leaves and white aerial roots that become tendrils.
Growth structure: The plant grows up to 50 feet (15 m) high. It climbs with fleshy shoots as thick as a finger. In cultivation, it doesn't grow as tall.
Leaves: The leaves are alternating, almost sessile, tapering, and very smooth. They are 3.2 to 9.6 inches (8 to 24 cm) long and .8 to 3.2 inches (2 to 8 cm) wide. The veins are hard to recognize.
Flowers: The flowers grow in short racemes that are like simple umbels. The petals are pale greenish to light yellow. They are 1.6 to 2.8 inches (4 to 7 cm) long. The inner petal is shorter and rolled together into a tube with a wavy nozzle. The outer petal is usually attached to this tube. The flower is only open for a few hours in the morning.
Fruits: The fruits are oblong and hang. They are 4 to 12 inches (10 to 30 cm) long and .2 to .6 inch (0.5 to 1.5 cm) thick. The fruits are yellowish when they are ripe. Inside are tens of thousands of tiny seeds that look like little black grains.
Occurrence: Vanilla originated in Mexico, but, today, it is cultivated in many tropical areas, including Madagascar, the Comoro Archipelago, and Réunion.
Other names: German "Vanille"; Spanish "Vainilla."
Additional information: Real vanilla is the second most expensive spice in the world. (Saffron is the most expensive.) This is not surprising, considering how expensive it is to produce. Except for plants growing in Mexico, each flower must be pollinated by hand because the special bee that pollinates this plant is only found in Mexico. It takes five to eight months before the fruits are completely ripe; any sudden onset of cold weather in this period destroys the entire harvest. After the harvest, the fruits are heated and spread out in the sun daily for three to six months. Then, they are wrapped in woolen blankets to ferment. In order to preserve their distinctive aroma, they are kept in airtight containers overnight.

Pepper

Piper nigrum
Family: *Piperaceae*, Pepper

Most important features: The stem has clinging aerial roots at the bulged knots. The lateral veins of the leaves run from the base to the tip.
Growth structure: Pepper grows up to 50 feet (15 m) high as a climbing plant. In cultivation, it seldom reaches this height. It often grows in a zigzag manner and is woody at the ground.
Leaves: The leaves are alternating, ovate to elliptical, bare, and pointed. They are 3.2 to 8 inches (8 to 20 cm) long and 2 to 6 inches (5 to 15 cm) wide. The leaves often have two pairs of lateral veins from the bottom and one pair slightly above it. These are not exactly opposite each other.
Flowers: The flowers are tiny and insignificant. They grow in dense hanging spikes that are 1.2 to 8 inches (3 to 20 cm) long.
Fruits: The fruits are spherical and .2 inch (5 mm) long. They have a large pit and are red when ripe.
Occurrence: Pepper is cultivated in many humid tropical areas. Originally, it was from the southwestern part of India.
Other names: German "Pfeffer"; French "Poivre"; Portuguese "Pimenta"; Spanish "Pimienta."
Additional information: Pepper is the most important spice. It has been used in Europe since Roman times. Black pepper is created by fermenting the fruits before they are ripe. If they are rapidly dried or marinated, they remain green. White pepper is made from the pits of the ripened picked fruits. Red pepper is made from dried ripe fruits and is often confused with the fruits of the pepper tree (see page 74). Several of the approximately two thousand pepper species are still in cultivation. The fruits of the African pepper (*P. guineense*), the Indian long pepper (*P. longum*), and the cubeb pepper (*P.cubeba*) are used. However, only the leaves of the betel pepper (*P. betle*, compare with page 36) are used. In Polynesia, an intoxicating drink is prepared from the roots of the kava (*P. methysticum*).

Cocoyam

Colocasia esculenta
Family: *Araceae*, Arales

Most important features: The leaves are large, ovate, and incised at the bottom half way to the stipula.
Growth structure: Cocoyam is an herbaceous plant with a tuberous rootstock. Usually, only the leaves emerge from the ground or water.
Leaves: The leaves grow on firm, erect petioles that are 8 to 60 inches (20 to 150 cm) long. The leaf blades are 5.2 to 24 inches (13 to 60 cm) long and 4 to 15 inches (10 to 35 cm) wide. They hang down and often are slightly wavy. The basal lobes are spherical at the end.
Flowers: The flowers are tiny and grow on cobs that are 1.6 to 8 inches (4 to 20 cm) long. Three to six cobs are enclosed in a bract that is cream white to yellow. The bracts are 3.6 to 16 inches (9 to 40 cm) long and .8 to 2 inches (2 to 5 cm) wide. One seldom sees the flowers in cultivated plants.
Fruits: The fruits are green, shiny berries that are dense on the cob. They are up to .2 inch (5 mm) long.
Occurrence: The cocoyam is cultivated on swampy fields. When it grows wild, it usually grows in ditches. It originated in Asia.
Other names: English "Dasheen," "Eddoe," "Elephant's Ear," "Kalo," "Yam"; German "Taro"; Portuguese "Inhame"; Spanish "Chamol," "Malanga," "Ñampi," "Ocumo Chino."
Additional information: The cocoyam has been cultivated in Asia for thousands of years. In Polynesia, it is still considered a basic food. The tubercle has distinct, ring-shaped leaf scars. It contains a lot of starch and oxalic acid. It is cooked by boiling it or roasting it. There are many specimens in cultivation. These have dark olive green to very pale or even somewhat spotted leaves. *Dioscorea* species are also called "yams." They are climbing plants with heart-shaped leaves.

Sweet Potato

Ipomoea batatas
Family: *Convolvulaceae*, Creepers

Most important features: The stem is crawling and roots at the tubercles. The leaves are heart-shaped at the bottom. Usually, they are slightly wavy between the veins. The flowers are relatively large.
Growth structure: The stem grows up to 16 feet (5 m) long. It is strongly branched and often slightly fleshy.
Leaves: The leaves are alternating and grow on long petioles. The leaves have a variety of shapes, ranging from ovate to deeply palmately cleft. They have three to seven lobes. The leaves are 1.6 to 6 inches (4 to 15 cm) long and 1.2 to 4.4 inches (3 to 11 cm) wide.
Flowers: The flowers usually grow in small groups. The sepals are tapering and about .4 inch (1 cm) long. The outer ones are acute and often shorter than the inner ones. The corolla is shaped like a bell or funnel. It is lavender blue to red violet with a darker throat (occasionally white). The corolla is 1.2 to 2.8 inches (3 to 7 cm) long. Some cultivated specimens are entirely without flowers.
Fruits: The fruits are ovate with a few large, dark seeds. They rarely develop.
Occurrence: The sweet potato is cultivated in all tropical areas. Originally, it was from the Western Hemisphere.
Other names: German "Süßkartoffel," "Batate"; French "Patate Douce"; Spanish "Batata," "Boniato," "Camote."
Additional information: The sweet potato belongs to the most important group of useful tropical plants. There are numerous cultural specimens with leaves and tubercles in a variety of shapes and colors. The tubercles contain starch, but they also have some sugar, which gives them a sweet taste. They are used as food, as starch, and as the basis for alcohol. The water spinach (*I.aquatica*), a swamp plant with triangular leaves that plays an important role in the cuisine of Southeast Asia, is similar.

Peanut

Arachis hypogaea
Family: *Fabaceae*, Papilionaceous plants

Most important features: The leaves are bipinnate. The flowers are yellow and shaped like peas. Later on, they are anchored in the earth.
Growth structure: This is an annual herbage that is richly branched. The stem is squared and often lies on the ground. The plant grows 6 to 28 inches (15 to 70 cm) high.
Leaves: The leaves are alternating. At the bottom, they are slender on stipules that are .8 to 1.2 inch (2 to 3 cm) long. The midrib is 1.2 to 4.8 inches (3 to 12 cm) long. The pinnulas are elliptical and spherical at the end. They are .8 to 2.4 inches (2 to 6 cm) long and .6 to 1.2 inch (1.5 to 3 cm) wide.
Flowers: The flowers grow individually or in groups of up to six close to the stem. The calyx is 1 to 2 inches (2.5 to 5 cm) long with four dentate labrum and an undivided labium. The five petals are .4 to .6 inch (1 to 1.5 cm) long. One of them is bent upward; two are sideways; and two are stuck together in a boat-shaped formation.
Fruits: The fruits are yellow brown, irregular, and spherical to tapering with netlike ribs. They are .8 to 2.4 inches (2 to 6 cm) long and .4 to .6 inch (1 to 1.5 cm) thick. They have one to four seeds inside a thin red brown peel. The fruits ripen underground.
Occurrence: One finds peanuts cultivated in large fields in the tropics and subtropics. Originally, they were from the southeastern part of South America.
Other names: English "Ground Nut"; German "Erdnuss"; French "Arachide"; Portuguese "Amendoim"; Spanish "Cacahuete," "Maní."
Additional information: The relatively short-lived peanut reserves its spot for the next generation by planting its seeds itself. After flowering, it stretches the base of the ovary up to 8 inches (20 cm) so that the front part, which encompasses the ovule, is pressed underground. The fruit ripens there. Peanuts are rich in fat and protein. They are used in a multitude of ways.

Soybean

Glycine max
Family: *Fabaceae*, Papilionaceous plants

Most important features: All parts, except for the petals, have stiff, bristly hair. The leaves have three to five pinnules. The flowers are shaped like peas. They are white, bluish, or violet. The fruits are shaped like beans.
Growth structure: This is an annual herbage that grows erect up to 8 to 40 inches (20 to 100 cm) high. On occasion, it reaches 7 feet (2 m).
Leaves: The leaves are alternating. They are 2.8 to 8 inches (7 to 20 cm) long. The stipules are .2 inch (5 mm) long. The pinnulas are ovate. They are 1.2 to 6 inches (3 to 15 cm) long and .8 to 4 inches (2 to 10 cm) wide. Usually, they are spherical at the bottom.
Flowers: Three to eight flowers grow in simple racemes that are .4 to 1.6 inch (1 to 4 cm) long. The calyx is .2 to .3 inch (5 to 8 mm) long. The two upper sepals are connected more than the three lower ones. The flowers have five petals. The upper one is spherical and .2 inch (5 mm) long. The two on the sides are smaller. The two lower ones are stuck together in the shape of a boat that contains the stamens and the pistil.
Fruits: The fruits are tapering. They are 1.2 to 3.2 inches (3 to 8 cm) long and .3 to .6 inches (8 to 15 mm) thick. When ripe, they are brownish yellow with two to four seeds.
Occurrence: Originally from Asia, soybeans are often cultivated in large fields. They are grown more frequently in the subtropics than in the tropics.
Other names: In most European languages "Soja" or "Soya"; German "Sojabohne."
Additional information: The soybean is one of the most important plants in the world. Its seeds are about twenty percent carbohydrates, thirty percent oil, and forty percent protein. This is roughly the same percentage as that found in animals. Thus, soybeans are an important food in feeding animals. Many ready-to-serve meals contain soy. Tofu is made of soy protein. It plays an important role in Asian cuisine. Soybean oil can be used for cooking; it can also be used in industry. Even artificial hormones are produced from soybeans.

Cassava

Manihot esculenta
Family: *Euphorbiaceae*, Wolf's milk plants

Most important features: All parts of the plant contain a milky sap. The leaves have three to nine lobes. The male flowers are stunted; the female ones are pale green on the outside and partly crimson on the inside.
Growth structure: Cassava is a slightly woody bush that grows 3 to 16 feet (1 to 5 m) high. It has distinct leaf scars at the stem, but few branches.
Leaves: The leaves are alternating on petioles that are 2 to 6.8 inches (5 to 17 cm) long. The lobes are tapering. They are often widest slightly above the middle lobe. There, they are 2.4 to 6.8 inches (6 to 17 cm) long and .4 to 2 inches (1 to 5 cm) wide. On the bottom, they are usually waxy and bluish.
Flowers: The flowers grow in panicles at the ends of the stems. They are bell-shaped and .3 inch (8 mm) long. On the bottom of the inside is an orange nectar gland. Often, there are many male flowers, each of which has ten stamens. There are only a few female ones. These have an ovary and three fringed stigmatic branches.
Fruits: The fruits are spherical to elliptical with wavy longitudinal edges. They are .6 inch (15 mm) long. When the fruits are ripe, they burst into three parts.
Occurrence: Cassava is cultivated everywhere in the tropics. Originally, it was from the Western Hemisphere.
Other names: English "Tapioca"; German "Maniok"; Portuguese "Mandioca"; Spanish "Yuca."
Additional information: Cassava is a basic food for more than 500 million people. That makes it one of the most important nutritional plants in the world. Its underground tubercles are 12 to 35 inches (30 to 90 cm) long and 2 to 4 inches (5 to 10 cm) thick. They contain a lot of starch, but they also contain hydrocyanic acid that must be destroyed by washing it out or by heating it. One cannot imagine the Brazilian farinha nor the West African gari cuisine without cassava. Latex gloves are made from the milky sap of *M. glaziovii*, a related species.

Sugarcane

Saccharum officinarum
Family: *Poaceae*, Grasses

Most important features: Sugarcane is very large grass. The stem is unbranched. The top is covered by overlapping leaf sheathes; the rest of the stem is naked and waxy at the bottom. The nodes are up to 8 inches (20 cm) apart, but they are closer together toward the ground. The leaves have a light midrib and a sharp edge.
Growth structure: The plant grows erect. It is 7 to 26 feet (2 to 8 m) high and .8 to 2.8 inches (2 to 7 cm) thick. The stems are often yellowish or reddish with ring-shaped leaf scars above the root systems.
Leaves: The tapered leaves grow in two vertical rows. They are 20 to 80 inches (50 to 200 cm) long and 1.6 to 4 inches (4 to 10 cm) wide.
Flowers: The flowers are tiny and grow on spikes enclosed in white hair. Hundreds of them grow in featherlike panicles that are 12 to 40 inches (30 to 100 cm) high.
Fruits: The fruits are tapering grass grains; however, they are almost never seen with cultivated plants.
Occurrence: Sugarcane is cultivated everywhere in the tropics, including gardens and huge rocks. New Guinea is believed to be the place of origin.
Other names: German "Zuckerrohr"; French "Canne à Sucre"; Portuguese "Cana de Açúcar"; Spanish "Caña de Azúcar."
Additional information: More than half of the worldwide production of sugar comes from sugarcane. It collects in the core of the stems where it can account for twenty percent of the stem. Brown cane sugar comes from a coarse cleaning of the pressed juice. White sugar requires additional crystallizing. Often, any juice that cannot be crystalized is fermented to make alcohol. In many countries, pieces of sugarcane are sold as inexpensive sweets.

Index

Photo Credits

Amro Lacus: 41 abv. rt., 49 abv. lt., 57 bl. lt., 59 bl. lt., 79 bl. rt., 81 abv., 83 abv. lt., 85 bl., 93 abv. rt., 95 abv. rt., 145 abv. lt., 177 bl., 189 bl. lt., 235 bl., 247 bl. lt., 253 abv., 257 bl., 271 bl.;
Bittmann: 7, 29 abv., 57 abv. lt., 57 abv. rt., 61 abv. rt., 61 bl., 63 abv., 67 abv., 69 abv. lt., 75 abv., 89 abv., 95 abv. lt., 113 bl. rt., 121 bl., 135 bl., 161 bl., 185 bl., 199 bl., 221 bl., 241 bl.;
Crone: 227 abv.
Denzer: 67 bl. lt.
Dietrich: 14, 37 bl.rt., 43 bl. rt., 55 bl. rt., 69 abv. rt., 73 bl. rt., 111 bl. rt., 137 bl., 157 bl. lt., 163 bl., 237 abv. lt., 237 abv. rt., 249 bl.;
Dörr: 155 abv., 169 bl. rt., 179 abv., 193 abv., 241 abv.;
Dransfield J.: 47 abv. lt.;
Eisenreich: 2/3, 53 bl. lt., 59 abv. lt., 77 bl. rt., 99 abv. lt., 125 abv., 157 abv., 211 bl., 231 abv. lt., 237 bl. rt., 243 abv.;
Feuerer: 83 abv. rt., 87 bl. lt., 97 abv. rt., 107 abv. rt., 203 abv.;
Gerlach: 83 abv., 89 bl. lt., 99 abv., 113 bl. rt., 139 lt., 175 bl., 183 bl., 207 abv., 227 bl., 245 bl.;
Greissl: 101 abv. lt., 101 abv. rt., 109 bl. rt., 167 abv. lt., 197 abv., 231 abv. rt.;
Hagen: 39 abv. lt., 121 (Einklinker), 123 abv., 133 abv., 159 bl. rt., 169 abv., 197 bl., 225 abv. rt., 233 bl., 239 bl.;
Kögel: 19 lt., 143 abv.;
König: 31 abv. lt., 31 abv. lt., 35 bl. lt., 35 bl. rt., 43 bl. lt., 45 bl. rt., 61 abv. lt., 69 bl., 73 abv. lt., 85 abv. lt., 91 bl., 103 bl. rt., 113 abv. rt., 115 bl. lt., 117 bl., 137 abv., 139 abv., 143 bl., 145 bl. lt., 145 bl. rt, 147 abv., 149 bl. lt., 153 bl. lt., 205 bl. lt., 205 bl. rt., 207 bl. rt., 213 abv., 215 abv., 217 bl., 219 bl. lt., 223 bl., 229 bl., 233 abv., 243 bl. lt., 251 abv., 257 abv., 261 bl., 267 abv.;
Kriegr: 135 abv., 223 abv.
Lüpnitz: 41 bl. lt., 51 bl. rt., 55 abv. rt., 59 bl. rt., 79 (Einklinker), 91 abv., 129 abv. rt., 159 bl. lt., 161 abv., 171 bl., 191 bl.;
Morell: 53 abv. rt.
Niesler: 105 abv.
Nowak: 33 abv. lt., 33 abv. rt., 35 abv. rt., 37 abv. lt., 37 bl. lt., 39 bl. lt., 45 abv., 51 abv. rt., 53 abv. rt., 71 bl. lt., 71 bl. rt., 87 abv. rt., 101 bl., 107 bl. lt., 107 bl. rt., 111 abv. rt., 117 abv. lt., 119 abv., 209 abv., 255 bl.;
Pott: 33 bl. lt., 33 bl. rt., 51 abv. lt., 203 bl., 259 abv. lt., 269 abv. rt., 275 bl.;
Reinhard: 1, 20, 29 bl., 31 bl. rt., 37 abv. rt., 47 abv. lt., 47 bl. rt., 49 abv. rt., 49 bl., 53 abv. lt., 53 abv. rt., 59 abv. rt., 77 bl. lt., 97 bl., 109 abv., 115 bl. rt., 121 abv., 123 abv., 127 abv. lt., 127 bl., 129 abv. lt., 133 bl., 145 abv. rt., 147 bl., 149 abv., 149 bl. rt., 151 abv., 151 bl., 153 bl. rt., 155 bl., 159 abv., 167 bl., 169 bl. lt., 171 abv., 177 abv., 181 abv., 183 abv., 185 abv., 187 abv., 187 bl. rt., 189 abv., 191 abv., 193 bl., 199 abv., 205 abv., 209 bl., 211 abv. rt., 213 bl., 217 abv., 225 abv. lt., 225 bl., 229 abv. rt., 237 bl. lt., 243 abv. rt., 245 abv., 251 bl. lt., 253 bl. rt., 255 abv. lt., 261 abv., 265 bl., 269 abv. lt., 269 bl., 271 abv., 273 bl., 275 abv.;
Rohwer: 9 abv., 9 bl., 10, 11, 12, 16 lt., 16 rt., 18, 19 rt., 21, 31 abv. rt., 39 bl. rt., 41 abv. lt., 45 bl. lt., 47 bl. lt., 55 abv. lt., 55 bl. lt., 63 bl., 65 abv., 65 (Einklinker), 65 bl., 67 bl. rt., 71 abv. lt., 71 abv. rt., 73 abv. rt., 73 bl. lt., 75 bl. lt., 75 bl. rt., 79 abv., 79 bl. lt., 81 bl., 85 abv. rt., 87 abv. lt., 87 bl. rt., 93 bl. lt., 93 bl. rt., 95 bl., 97 abv. lt., 99 abv. rt., 103 abv., 105 bl., 107 abv. lt., 109 bl. lt., 111 abv. lt., 111 bl. lt., 113 abv. lt., 115 abv., 117 abv. rt., 119 bl., 125 bl. lt., 125 bl. rt., 127 abv. rt., 129 bl., 131 abv., 131 bl., 141 bl., 153 abv., 157 bl. rt., 163 abv., 165 abv., 165 bl., 167 abv. rt., 173 abv. lt., 173 abv. rt., 175 abv., 179 bl., 181 bl., 187 bl. lt., 189 bl. rt., 195 abv., 195 bl., 201 abv., 201 bl., 207 bl. lt., 215 bl., 219 abv., 219 bl., rt., 221 abv., 229 abv. lt., 231 bl., 235 abv., 239 abv., 247 abv., 249 abv. lt., 255 abv. rt., 259 abv. rt., 259 bl. lt., 259 bl. rt., 263 abv., 263 bl., 265 abv., 267 bl., 273 abv.;
Senghas: 247 bl. rt., 249 abv. rt., 251 bl. rt.;
Staeck: 57 bl. rt., 77 abv., 85 (Einklinker), 89 bl. rt., 93 abv. lt., 103 bl. lt., 173 bl., 211 abv. lt;
Thiv: 51 bl. lt;
Weber: 41 bl. rt., 141 abv., 253 bl. lt.:
Zona: 31 bl. rt., 39 abv. rt., 43 abv. lt;

Pictogram:
Graphics: See pg. 26/27; according to author's models

Library of Congress Cataloging-in-Publication Data Available

Translated by Nicole Franke and Daniel Shea

10 9 8 7 6 5 4 3 2 1

Published by Sterling Publishing Company, Inc.
387 Park Avenue South, New York, N.Y. 10016
Originally published in German under the title *Pflanzen der Tropen* and © 2000 by BLV Verlagsgesellschaft mbH, München

Distributed in Canada by Sterling Publishing
c/o Canadian Manda Group, One Atlantic Avenue, Suite 105, Toronto, Ontario, Canada M6K 3E7
Distributed in Great Britain and Europe by Cassell PLC
Wellington House, 125 Strand, London WC2R 0BB, England
Distributed in Australia by Capricorn Link (Australia) Pty. Ltd.
P.O. Box 704, Windsor, NSW 2756 Australia

Printed in Germany

Sterling ISBN 0-8069-8387-6